Understanding Commanders' Information Needs for Influence Operations

Eric V. Larson, Richard E. Darilek, Dalia Dassa Kaye,

Forrest E. Morgan, Brian Nichiporuk, Diana Dunham-Scott,

Cathryn Quantic Thurston, Kristin J. Leuschner

Prepared for the United States Army

Approved for public release; distribution unlimited

ARROYO CENTER

The research described in this report was sponsored by the United States Army under Contract No. W74V8H-06-C-0001.

Library of Congress Cataloging-in-Publication Data

Understanding commanders' information needs for influence operations /
    Eric V. Larson ... [et al.].
        p. cm.
    Includes bibliographical references.
    ISBN 978-0-8330-4691-8 (pbk. : alk. paper)
    1. United States—Armed Forces—Officers—Information services.
    2. Generals—United States—Information services. 3. Command of troops.
    4. Influence (Psychology) 5. Information warfare—United States. 6. Combined
    operations (Military science) 7. United States—Armed Forces—Information
    services. 8. United States—Armed Forces—Planning. 9. United States—Military
    policy. I. Larson, Eric V. (Eric Victor), 1957–

    UB413.U434 2009
    355.4'1—dc22

                                                                    2009042183

Published 2009 by the RAND Corporation
1776 Main Street, P.O. Box 2138, Santa Monica, CA 90407-2138
1200 South Hayes Street, Arlington, VA 22202-5050
4570 Fifth Avenue, Suite 600, Pittsburgh, PA 15213-2665
RAND URL: http://www.rand.org/
To order RAND documents or to obtain additional information, contact
Distribution Services: Telephone: (310) 451-7002;
Fax: (310) 451-6915; Email: order@rand.org

Preface

This is the final report for a RAND Arroyo Center study called "Integrating Influence and Information Operations into Army Planning and Operations." The objective of this study was to help improve the effectiveness of combined arms operations by characterizing commanders' requirements for information on cultural and other "soft" factors (e.g., networks and hierarchies, norms, attitudes) and by developing practical ways for commanders to integrate influence activities into combined arms planning and assessment.

This research was sponsored by the U.S. Army Information Operations Proponent (USAIOP), Combined Arms Center, U.S. Army Training and Doctrine Command (TRADOC), Ft. Leavenworth, Kansas. It was completed in September 2006, and the final report was submitted for sponsor approval in August 2007. Some policies and practices could have changed between report submission and receipt of clearance for publication. The research was conducted in RAND Arroyo Center's Strategy, Doctrine, and Resources Program. RAND Arroyo Center, part of the RAND Corporation, is a federally funded research and development center sponsored by the United States Army. Please direct any comments concerning this research or requests for additional information to the principal investigator, Dr. Eric V. Larson, at 310-393-0411, extension 7467, or larson@rand.org.

The Project Unique Identification Code (PUIC) for the project that produced this document is ATFCR06031.

For more information on RAND Arroyo Center, contact the Director of Operations (telephone 310-393-0411, extension 6419; FAX 310-451-6952; email Marcy_Agmon@rand.org), or visit Arroyo's Web site at http://www.rand.org/ard/.

Contents

Preface iii
Figures ix
Tables xi
Summary xiii
Acknowledgments xxiii
Abbreviations xxvii

CHAPTER ONE

Introduction 1
Defining Terms 2
Study Tasks and Analytic Approach 3
Organization of This Monograph 5

CHAPTER TWO

Commanders' Information Needs for Influence Operations 7
Insights from Structured Conversations with Commanders 7
Insights from Recent Papers by Senior Commanders 13
   GEN Peter W. Chiarelli, Commander, 1st Cavalry Division 13
   MG David H. Petraeus, Commander, 101st Airborne Division
      (Air Assault) 14
   LTG Thomas F. Metz, Commander, III Corps, Coalition Joint
      Task Force–7, and Multi-National Corps–Iraq 16
   COL Ralph O. Baker, Commander, 2nd Brigade Combat Team,
      1st Armored Division 16
   Key Common Insights from Commanders' Papers 18

Insights from Case Study Analyses 19
  Bosnia 19
  Kosovo.. 21
  Afghanistan... 22
  Iraq 23
  Key Trends and Contrasts 26
Insights from the National Training Center................................. 27
Insights from 1st Information Operations Command.......................... 29
Observations from Unified Quest 2006...................................... 30
Insights from a Review of Doctrine, Tactics, Techniques, and
      Procedures, and Task Lists... 33
Chapter Conclusions... 37

CHAPTER THREE
Sources of Commanders' Information Needs 41
Commanders' Guidance 41
The Operating Environment and Information Domain 42
  The Battlefield Environment... 43
  The Threat Domain... 43
  The Information Domain.. 44
Resources Available to the Commander...................................... 52
Chapter Conclusions... 52

CHAPTER FOUR
Remaining Challenges.. 57
Vertical Coordination and Echelonment..................................... 57
Horizontal Coordination Across Areas of Operation......................... 59
Ensuring Continuity in Transitions 59
Overcoming Doctrinal Stovepiping of Information Operations 61

APPENDIXES
A. Identified Information Requirements for Influence
    Operations 65
B. Task List Analysis.. 71
C. A Metrics-Based Planning and Assessment Approach for
    Influence Operations 81

D. Assessment of Expected Utility Modeling for Influence
    Operations 107
E. Assessment of Social Network Analysis for Influence
    Operations 119

References 127

Figures

S.1.   Geospatially Oriented Aspects of the Information
        Domain. xviii
3.1.   Geospatially Oriented Aspects of the Information
        Domain. 45
B.1.   Basic Distribution of Tasks Related to IO and Influence
        Operations 76
B.2.   Distribution of Functions Related to IO and Influence
        Operations 77
B.3.   Detailed Distribution Tasks Related to IO and Influence
        Operations Tasks in the AUTL 78
C.1.   Flows of Information for Planning and Assessment
        System. 82
C.2.   Elements of Metrics. 85
C.3.   Overall Flow of Metrics-Based Planning and Assessment
        Methodology 85
C.4.   Example of IO and Other Tasks 90
C.5.   Example: Measure of Outcome for AO Objective 91
C.6.   Example: Measure of Effectiveness 92
C.7.   Example: Measure of Performance 93
C.8.   The Planning, Preparation, Execution, and Assessment
        Cycle 95
C.9.   Notional Data Flow Through Matrixes 101
C.10.   Organizational Backbone to Metrics-Based Planning
        and Assessment System. 105
D.1.   Illustrative Data for Expected Utility Forecast,
        August 2002. 113
E.1.   Illustration of Pajek Network Customization 123

Tables

B.1   Core, Supporting, and Related Functions of IO and
Influence Operation 74

Summary

There is growing recognition within the Army and joint world that recent U.S. military operations in Iraq and Afghanistan—including information operations (IO) and influence operations—have turned in large measure on an understanding of cultural and other "soft" factors. However, along with this recognition of the importance of these factors have come many questions, including: How do commanders view their requirements for "cultural preparation of the environment"? How can these sorts of factors be considered more systematically in planning and conducting operations?

The objective of our study was to help improve the effectiveness of combined arms and joint operations by characterizing commanders' requirements for information on cultural and other "soft" factors, and by developing practical ways for commanders to integrate influence activities into combined arms planning and assessment. The research entailed structured conversations with commanders and their staffs, a review of senior commanders' and other writings on IO and influence operations, an analysis of task lists, and an assessment of relevant data from the 1st Information Operations Command and the National Training Center.

In our usage, the term *information operations* is as defined by the Department of Defense (DoD):

> [t]he integrated employment of the core capabilities of electronic warfare, computer network operations, psychological operations, military deception, and operations security, in concert with specified supporting and related capabilities, to influence, disrupt, cor-

rupt or usurp adversarial human and automated decision making while protecting our own.

The term *influence operations* can generally be understood as synonymous with *strategic communication* (STRATCOMM), which is defined in Joint Publication 5-0 as

> [f]ocused United States Government efforts to understand and engage key audiences to create, strengthen, or preserve conditions favorable for the advancement of United States Government interests, policies, and objectives through the use of coordinated programs, plans, themes, messages, and products synchronized with the actions of all instruments of national power.

Put simply, influence operations engender communications and interactions that aim to inform and influence target audiences in concert with other kinetic and non-kinetic activities. Of the core IO capabilities, psychological operations (PSYOP) are the most pertinent to influence operations. To simplify our presentation, we generally use the collective term *influence operations* throughout this monograph.

Commanders' Information Needs for Influence Operations

Our review of a range of sources provided us with a number of insights into commanders' information requirements for influence operations. Perhaps the most important insight is that for the types of contingencies in which the U.S. Army now finds itself (counterinsurgency [COIN] and stability operations), the most critically needed information may have to do with understanding the attitudes, beliefs, and mood of the local civilian populations.

During recent operations, inadequate information on the attitudes and beliefs of local populations has often led to bland messages that did not resonate with specific target audiences and that made it difficult to compete with adversaries more capable of exploiting the local information environment. Understanding the popular mood requires continuous monitoring of key indicators, perhaps more so in Muslim societies that are innately suspicious of the West and the United States. Shifts in popular opinion are especially likely after a single traumatic incident, whether it is a bombing raid that causes severe collateral damage to civilian homes and property or a traffic accident in which U.S. military vehicles accidentally kill a local child.

Our research suggests, furthermore, that success in influence operations depends on commanders' views of the battle space, their understanding of how to employ influence operations to achieve desired end states, and their interest and involvement in integrating IO with other combined arms operations. Commanders who insist that their subordinates develop a coordinated program of IO and influence operations activities and who follow up to ensure these activities take place appear far more likely to succeed in integrating influence operations into the campaign than commanders who take a more passive view of influence operations. Commanders also need to reemphasize the importance of influence operations on a regular basis.

Our research also revealed that there is no single correct answer to the question of which sources of information ought to be drawn upon to accurately assess the local information environment. The most appropriate sources will vary according to the mission, the local context of the operation, and even the individual commander. It is important, however, to establish a clear information sourcing strategy in an area of operation (AO) very early, so that subordinate commanders know what is expected of them over the long term.

We also found that commanders who believed their influence operations had been successful invariably had a clear, uncluttered picture of the key influence variables in the current battle space, the resources available to support influence operations, and the end state they desired for the end of the tour of duty. Commanders who tried to monitor too many variables, who shifted resources back and forth in response to daily crises without a long-term steady state, or who changed themes and messages randomly without any underlying concept of a step-by-step path to victory—these commanders appear to have enjoyed less success.

Developing good measures of effectiveness (MOEs) to assess how a unit's influence efforts are being received by the local population is one of the thorniest problems facing the Army today. Although none of our interlocutors believed that the Army has a particularly good set of MOEs for influence operations in COIN and stability operations, our interviews revealed that three key indicators in particular are being used across units and echelons in Iraq and Afghanistan with some success: the tenor of sermons in mosques, the "on the street" behavior of the locals (obscene gestures toward U.S. troops, amount of anti-American graffiti, etc.), and trends, either upward or downward, in the number of intelligence tips from the local population.

A Framework for Thinking About Commanders' Information Requirements

Our conversations with commanders and our review of the written record suggest that commanders' needs for information generally flow from an interaction of factors within three principal arenas: commanders' guidance regarding the overall mission; the resources available to the commander, which are likely to vary from operation to operation and over time; and the operating environment, including the information domain.

In terms of commanders' guidance, influence operations planning should flow from the top down and be designed and executed in support of coherent politico-military objectives while simultaneously synchronizing and/or integrating kinetic and non-kinetic activities, whether they are conducted by the services or by other DoD or interagency actors. Importantly, units in the field also need the authority and flexibility to operate within these broader, higher-level parameters if they are to be responsive to quickly developing opportunities and challenges. Satisfactorily resolving the tensions between these two desiderata appears to be key to success.

Beyond understanding the forces and other resources under his immediate command, a commander must understand the forces and other resources available under the command of higher echelons or in adjacent AOs that may impact his operations, those being assigned to him, and those assigned to subordinate commanders.

The operating environment arena, especially the information domain, is more complex and requires more discussion. There are currently a great many terms and phrases in use that attempt to capture the most salient features of the contemporary operating environment (e.g., "complex environments," "cultural environment," "cultural intelligence," "cultural preparation of the environment"), but there is little agreement on which framework or terminology should be used, or about exactly what the different terms mean. Our study provides what we believe is a fairly complete and highly intuitive framework for thinking about commanders' information requirements in COIN and stability operations, and for guiding data collection efforts related to the information domain. Moreover, the endorsements we received from commanders and members of battle staffs who were presented with this framework suggest its potential utility as an organizing principle for system and database development.

Lenses to Characterize and Diagnose Features of the Information Domain

Given that the sorts of data and intelligence that are most important to commanders in any given operation are quite context specific and are influenced by the mission, commander, and various other factors, our framework uses three complementary "lenses" to characterize and diagnose features of the operating environment's information domain that are likely to constrain the effectiveness of influence operations and mission performance. Each lens focuses on one kind of information: (1) geospatial, (2) network oriented, or (3) tied to specific political or military stakeholder groups or their leaders.

Geospatially Oriented Information. The geospatial lens for understanding commanders' information needs captures a number of critically important features of the information domain that were identified in our interviews and literature reviews. Our research suggests that many characteristics of the geospatial component of the information domain are best portrayed as a set of overlapping layers, as shown in Figure S.1.

Figure S.1
Geospatially Oriented Aspects of the Information Domain

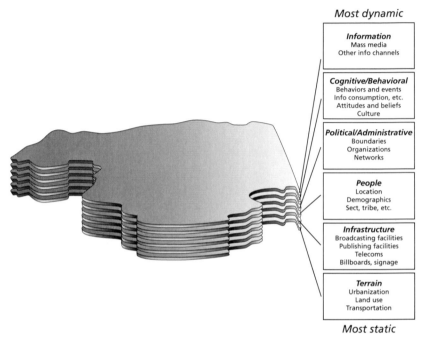

Most dynamic

Information
Mass media
Other info channels

Cognitive/Behavioral
Behaviors and events
Info consumption, etc.
Attitudes and beliefs
Culture

Political/Administrative
Boundaries
Organizations
Networks

People
Location
Demographics
Sect, tribe, etc.

Infrastructure
Broadcasting facilities
Publishing facilities
Telecoms
Billboards, signage

Terrain
Urbanization
Land use
Transportation

Most static

RAND *MG656-S.1*

These layers, or levels, range from mostly static features of the terrain (such as urbanization, land use, and transportation networks) to more-dynamic features of the environment (such as the changing attitudes, beliefs, and behaviors of a given population in a specific region, and the ever-changing mix of new messages and information competing for attention at any given time).

Network-Oriented Information. A second lens for unpacking the information domain of the operating environment can be characterized as overlapping or interlocking networks. This lens provides a view of key features of the broader, political society, including key leaders, their critical relationships, and their sources of authority, power, and influence. Networks can be used to characterize a host of formal organizations or hierarchies—whether they are political, military, bureaucratic, or administrative; economic or business oriented; tribal, religious, or sectarian. Networks also can be used to characterize informal networks, including those that are personal and professional, or that characterize patronage relationships or criminal enterprises, jihadist discourse, or influence. In addition, physical networks—such as telecommunications; command, control, communications, and computers; and utilities—translate naturally into link and node data. Such data, although quite helpful for influence operations, also can be somewhat difficult to compile and maintain, however.

Political or Military Stakeholder Groups and Their Leaders. Another lens through which to understand information and influence operations is provided by target audience analysis. This process involves identifying which groups or individuals need to be targeted, and whether targeting them means informing, influencing, cultivating, or incapacitating them. Each group or faction needs to be characterized in terms of its group identity and general worldview, as well as its specific aims, grievances, motivations, intentions, morale, basic strategies, leadership, and organizational structure. It also may be necessary to collect and maintain a number of types of information on key individuals who influence developments and thus need to be directly or indirectly courted or influenced.

Remaining Challenges

In addition to producing findings that can help the Army and the joint world make progress in conducting effective influence operations, our research identified four emerging challenges that need to be addressed.

Ensuring Vertical Integration of Information and Influence Operations Across Echelons

A recurring theme from our research is the need for integrated planning, execution, assessment, and information flows between echelons to ensure complementarity and synergy in influence operations. Brigade- and battalion-level personnel noted emerging difficulties between brigade- and corps-level influence operations, as well as between brigade and battalion operations. Commanders and former officers on battle staffs suggested that perhaps the biggest challenges lie in the battalion-brigade relationship, where disconnects between themes and messages and long approval times appear to be especially significant. The question of how best to balance the dual desiderata of top-down strategic and operational guidance with tactical-level authority and flexibility in execution to ensure responsiveness seems likely to be a recurring challenge for future commanders. Vertical integration might be enhanced by the adoption of the sort of top-down, metrics-based planning and bottom-up assessment process we describe.

Ensuring Horizontal Coordination and Integration Across Adjacent Areas of Operation

Our interviews and other research suggest that the importance commanders place on coordinating influence operations activities with commanders in adjacent AOs, and the mechanisms used to ensure this coordination, are somewhat ad hoc in nature. Our structured conversations with commanders and former members of battle staffs indicate that difficulties in synchronizing across AOs have led to different messages being emphasized at different times in different sectors. Such practices may result in confusion among Iraqis who move across brigade boundaries or talk to relatives in other AOs and find that different messages are being emphasized, and may raise questions about what the principal U.S. message might be at any given time. While the metrics-based planning and assessment process we describe could enhance the transparency of activities conducted by units in adjacent AOs, so, too, could other, less formal communications between units.

Ensuring Continuity in Information and Influence Operations Across Rotations

It is also critically important that newly arriving commanders be cognizant of and honor the promises made by their predecessors and minimize abrupt changes in influence operations that may confuse or increase uncertainty or fear among locals. In particular, significant efforts should be made to ensure both greater continuity in the application of influence operations across brigade rotations and the availability of operation-relevant information about the information domain and local population across rotations.

Our interviews with commanders suggest that current efforts to ensure smooth transitions between units, and to thereby enhance a sense of continuity in influence operations, may be inadequate. Some commanders thought that rather than building upon lessons learned by their predecessors during earlier unit rotations in an AO, units have tended to rotate in and immediately begin making changes without making a full appraisal of what elements of IO and influence operations might already be working. Additional efforts and mechanisms are needed to provide units that are rotating in with an endowment of relevant experiential information—chronologies, network analyses, contact files, databases, and other types of information—that can assist a new commander in understanding the history and authority structures of the AO. Moreover, it is not clear that the incentives for commanders support the sort of continuity needed for effective influence operations: Rather than rewarding a commander for making changes to his predecessor's influence operations, it might be better to reward him for improvements in relevant metrics. A common system and database not only could enable deploying units to monitor developments in the AO into which they will be deploying, but also could foster the development of institutional memory needed to achieve the desired level of continuity.

Overcoming Doctrinal Stovepipes

The final challenge is what we see as a necessary doctrinal shift, moving from a joint and Army conception of influence operations as a set of discrete stovepipes to one that focuses more on their contributions to achieving the objectives of combined arms, joint, and combined operations.

Our interviews and other analyses suggest that the success of influence operations in the field increasingly depends on commanders' ability to think beyond current doctrine, which tends to focus on the employment of IO in major combat operations, and treats IO and its related and supporting capabilities as discrete, somewhat isolated disciplines rather than capabilities whose employment needs to be planned, synchronized, and executed in concert with the other combined arms to produce desired effects and outcomes.

We think that Army influence operations doctrine should better consider the employment of influence operations across a wider range of operation types, from COIN and stability operations to major combat operations, and should focus more on principles for effectively integrating influence operations disciplines with traditional combined arms. Additionally, Army education and training should train future commanders in the principles of employing influence operations across a wider range of mission types, and should seek to promulgate best practices from the field for better integration of influence operations into combined arms operations.

In combination, these measures, if coupled with educational and training programs that teach soldiers how to integrate influence efforts with other activities, could give the next generation of Army commanders the tools they need to plan and execute more-effective influence operations.

Acknowledgments

This work on the rapidly changing field of information operations (IO) and influence operations would simply not have been possible without the generous assistance of a number of insightful and energetic Army officers and civilian analysts who are at the cutting edge of this discipline. Our RAND study team is very grateful to all of them.

We would like to thank Mr. Richard Kaplan and Dr. Joe Clema of the 1st Information Operations Command at Fort Belvoir, Virginia, who were kind enough to give us unfettered access to the command's archival records of IO reachback requests from units in Iraq and Afghanistan and also cheerfully arranged for us to interview several of the command's top analysts.

Our team's participation in the U.S. Army Training and Doctrine Command/Joint Forces Command (TRADOC/JFCOM) Unified Quest 2006 wargame at the Army War College in April 2006 was invaluable in that it allowed us to observe how influence operations are actually integrated into stability and support operations campaign planning. This participation was the result of efforts by COL Bob Johnson of the TRADOC Army Capabilities Integration Center, who permitted us to observe the different teams without restriction during the course of the game. He also allowed us to submit our early insights on the IO part of the game to the exercise control cell.

At the U.S. Army's National Training Center (NTC) at Fort Irwin, California, CPT Keith Wilson, one of the NTC's IO observer-controllers, met with members of the project team twice to discuss the different ways in which the various training units have approached IO at the NTC. Captain Wilson also provided us with some excellent raw data on IO plans and employment during several recent training rotations. We very much appreciate his efforts.

In the Washington, D.C., area we interviewed several active duty and recently retired senior Army officers (O-6 and above) who had commanded brigade-, division-, and corps-level units in the Balkans, Afghanistan, and Iraq. Some of these officers now work on the Army and Joint Staffs in the Pentagon, and others are now at research and analysis institutions. Because we promised them anonymity because of potential sensitivities associated with recent and ongoing influence operations, we cannot reveal their names; but suffice it to say that the insights these individuals provided could not have been obtained from any other readily accessible source. We are truly indebted to them for spending some of their valuable time with us. We also would like to thank Dr. Christopher Lamb of the National Defense University, who took considerable time to give us his impressions of the current state of the U.S. Army's psychological operations force structure and capabilities.

We concluded our research with a second trip to the Army War College, where we interviewed a wide selection of O-5 and O-6 level officers in the incoming Class of 2007 who have battalion command experience in Afghanistan and Iraq. We are very grateful to LTC Keith Pickens of the College's Operations Directorate for setting up our interview schedule on short notice with great skill and enthusiasm. In this instance as well, we promised anonymity to our interlocutors; but we thank them all for their open, honest, and detailed assessments of the strengths and weaknesses of influence operations as the U.S. Army employs them today. These interviews greatly enriched our research. We also had the good fortune to meet with two of the College's faculty members, COL John R. Martin (U.S. Army, ret.) and Professor Dennis Murphy, both of whom offered compelling insights into the strengths and weaknesses of current U.S. military doctrine for information and influence operations.

We wish to thank the following individuals for their insights into the application of the expected utility model forecasting approach as a tool for supporting influence operations: COL Wilfredo Colon, U.S.

Southern Command, COL Kevin Hendrick, U.S. Pacific Command, and Ed Smith of the Institute for Defense Analyses.

We would like to thank COL Rod Turner and LTC Chuck Eassa of the U.S. Army Information Operations Proponent's Office in the Combined Arms Center at Fort Leavenworth, Kansas, for the opportunity to continue our work in this area in fiscal year 2006.

We also would like to express our gratitude to Dr. Dorothy Denning of the Naval Postgraduate School and Dr. Walt Perry of RAND for their careful and helpful reviews of an earlier, draft version of this report. We also acknowledge an intellectual debt to—and absolve of any responsibility for the present report—our colleague James P. Kahan, whose 1989 study *Understanding Commanders' Information Needs* provided not only a creative and timeless way of thinking about the subject of commanders' information needs, but also the first part of the title of the present report.

Finally, we would like to thank RAND colleagues Lauri Zeman and Tom Szayna, program director and associate program director, respectively, of RAND Arroyo Center's Strategy, Doctrine, and Resources program, for their support during the course of the study, and Theresa DiMaggio and Martha Friese for their invaluable administrative assistance.

Abbreviations

1st IOC	1st Information Operations Command
ACIC	Army Counterintelligence Center
AIF	anti-Iraqi force
AO	area of operation
AOR	area of responsibility
ARCIC	U.S. Army Capabilities Integration Center
ASCC	Army Service Component Command
ASCOPE	areas, social structure, culture, opportunity, power and authority, economy
AUTL	Army Universal Task List
BCT	brigade combat team
C2	command and control
C3I	command, control, communications, and intelligence
CAC	Combined Arms Center
CALL	Center for Army Lessons Learned
CCIR	Commander's Critical Information Requirement
CD	compact disk
CFC	Coalition Force Command
CFFI	critical friendly force information
CFLCC	Coalition Forces Land Component Command
CIA	Central Intelligence Agency
CJCSM	Chairman of the Joint Chiefs of Staff Manual
CMO	civil-military operations
CNA	computer network attack

CND	computer network defense
CNE	computer network exploitation
CNO	computer network operations
COA	course of action
COE	contemporary operating environment
COIN	counterinsurgency
DIME	diplomatic, information, military, economic
DoD	Department of Defense
DPEG	demographic, political, economic, geographical
DVD	digital video disk
EEFI	essential elements of friendly information
FFIR	friendly force information requirements
FM	Field Manual
FMI	Field Manual Interim
FY	fiscal year
HBCT	heavy brigade combat team
HPT	high-priority target
HUMINT	human intelligence
HVT	high-value target
IED	improvised explosive device
IO	information operations
IPB	intelligence preparation of the battlefield
ISR	intelligence, surveillance, and reconnaissance
JFC	Joint Forces Commander
JFCOM	Joint Forces Command
JFLCC	Joint Force Land Component Commander
JP	Joint Publication
JTF	Joint Task Force
LOO	line of operation
MDMP	military decisionmaking process
METT-TC	mission, enemy, terrain and weather troops available, and civilian considerations
MNF	multi-national forces
MOE	measure of effectiveness

MOO	measure of outcome
MOP	measure of performance
NATO	North Atlantic Treaty Organization
NGIC	National Ground Intelligence Center
NGO	nongovernmental organization
NTC	National Training Center
OIF	Operation Iraqi Freedom
PIR	priority intelligence requirement
PMESII	political, military, economic, social, infrastructure, information
PSYOP	psychological operations
SASO	stability and support operations
SBCT	Stryker brigade combat team
SIAM	Situational Influence Assessment Model
SNA	social network analysis
TRADOC	U.S. Army Training and Doctrine Command
TTP	tactics, techniques, and procedures
UJTL	Universal Joint Task List
UQ 06	Unified Quest 2006
VBIED	vehicle-borne improvised explosive device

CHAPTER ONE

Introduction

The U.S. Army and joint world is increasingly recognizing that recent U.S. military operations in Iraq and Afghanistan—including information and influence operations—have turned in large measure on an understanding of cultural and other "soft" factors. And along with this recognition have come many questions about these factors and their role in military operations. How do commanders view their requirements for "cultural preparation of the environment"? How can these sorts of factors be considered more systematically in planning and conducting operations? How can influence operations become a more integrated part of the combined arms team? How can they be more useful tools for commanders? How can commanders and their staffs conceptualize and carry out better influence operations?

In fiscal year (FY) 2005, RAND Arroyo Center conducted a study on information operations (IO) in Operation Iraqi Freedom (OIF) for the U.S. Army's Combined Arms Center (CAC), and a study on influence operations for the U.S. Army Training and Doctrine Command (TRADOC) Futures Center (now the TRADOC Army Capabilities Integration Center, or ARCIC). The study for CAC surveyed the track record of IO activities in OIF, attempted to assess their contributions to operational-level outcomes, and developed a methodology for future planning and evaluation efforts that can improve the Army's ability to gauge the contribution of IO to the overall operational objectives of a military campaign. The study of influence operations for TRADOC assessed the potential value of a number of social science tools and methodologies that could enhance the capabilities available to commanders for assessing and addressing cultural, social, psychological, and other "soft" factors that can affect military operations.[1]

Defining Terms

It is important at the outset to define what we mean by *influence operations* and *information operations*. As used here, *influence operations* is an overarching term that subsumes or subordinates the capabilities of information operations and other activities to achieve influence objectives. Influence operations can generally be understood as synonymous with the term strategic communications (STRATCOMM), which Joint Publication (JP) 5-0 defines as

> [f]ocused United States Government efforts to understand and engage key audiences to create, strengthen, or preserve conditions favorable for the advancement of United States Government interests, policies, and objectives through the use of coordinated programs, plans, themes, messages, and products synchronized with the actions of all instruments of national power.[2]

Information operations is in turn defined as

> [t]he integrated employment of the core capabilities of electronic warfare, computer network operations, psychological operations,

---

[1]   See Eric V. Larson, Richard E. Darilek, Daniel Gibran, Brian Nichiporuk, Amy Richardson, Lowell Schwartz, and Cathryn Thurston, *Foundations of Effective Influence Operations: A Framework for Enhancing Army Capabilities*, MG-654-A, Santa Monica, Calif.: RAND Corporation, forthcoming.

[2]   Department of Defense (DoD), *Joint Operations Planning*, JP 5-0, Washington, D.C., December 26, 2006b, p. xii. For a more complete description of our usage of the term *influence operations*, see Larson et al., forthcoming. By comparison, the Air Force definition of *influence operations* as of January 2006, was: "Informing and appropriately influencing key audiences by synchronizing and integrating communication efforts to deliver truthful, timely, accurate, and credible information: Strategic refers to source of information, message, messenger, audience, timeframe, and/or effect; Communication refers to both what you say and what you do; Requires focus on both internal and external communication efforts; and Requires both peacetime and wartime processes and capabilities."

military deception, and operations security, in concert with specified supporting and related capabilities, to influence, disrupt, corrupt or usurp adversarial human and automated decision making while protecting our own.[3]

Of the core IO capabilities, psychological operations (PSYOP) are the most pertinent to influence operations.[4] To simplify our presentation, we generally use the collective term, *influence operations*, throughout the document.

Study Tasks and Analytic Approach

The two FY 2005 studies described above laid the foundation for a second study for CAC in FY 2006. This one focused on three major tasks:

- identifying commanders' information needs for conducting influence operations
- assessing the adaptability of select social science methodologies and tools to help meet commanders' needs for influence operations
- refining a metrics-based planning and assessment process developed in the earlier, FY 2005 work to make it suitable for Army employment.

The analytic approach the study team used was as follows.

The most resource-intensive part of our study was our effort to understand commanders' own views about their information and analytic needs for conducting influence operations, especially during

---

[3] DoD, *Department of Defense Dictionary of Military and Associated Terms*, JP 1-02, Washington, D.C., April 12, 2001a (as amended through October 17, 2008), p. 263.

[4] JP 1-02 (DoD, 2001a, p. 441) defines *psychological operations* as follows: "Planned operations to convey selected information and indicators to foreign audiences to influence their emotions, motives, objective reasoning, and ultimately the behavior of foreign governments, organizations, groups, and individuals. The purpose of psychological operations is to induce or reinforce foreign attitudes and behavior favorable to the originator's objectives."

counterinsurgency (COIN) and stability operations. For this task, we undertook the following analytic activities:

1. conducted structured conversations with corps-, division-, brigade-, and battalion-level commanders and their staffs to get their views on commanders' information needs for influence operations
2. reviewed Army commanders' own writings about IO and influence operations, as well as the sorts of information they considered crucial to the success of their operations
3. conducted case study analyses of lessons learned in Bosnia, Kosovo, Afghanistan, and Iraq related to commanders' information needs for influence operations
4. reviewed briefings, data, and other materials from brigade rotations at the National Training Center (NTC) at Fort Irwin, California, to identify relevant information collected by the brigades
5. reviewed Requests for Information made of the 1st Information Operations Command (1st IOC) to understand the information requirements from the field that are serviced by 1st IOC
6. reviewed relevant doctrinal publications; tactics, techniques, and procedures (TTP); training handbooks and guides; and other materials to understand what sorts of commanders' information needs might already have been identified in the doctrinal literature
7. observed the Unified Quest 2006 (UQ 06) exercise to better identify the sorts of information sought for political and military decisionmaking.

These efforts led to the development of a taxonomy of key information types and a framework for thinking about the different sorts of information needed for influence operations.

Earlier work for ARCIC identified the high potential utility of two key social science methodologies in the analysis and planning of influence operations—agent-based rational choice, or expected utility, models, and social network analysis (SNA) tools. The sponsor of the present study requested that we provide additional discussion of how these tools might be incorporated in Army, combined, and joint analysis and planning of influence operations. To address this task, we reviewed existing and emerging doctrine and discussed the use of these tools with representatives of the National Defense University, the Joint Warfare Analysis Center, the Army Science Board, the National Ground Intelligence Center (NGIC), and the regional combatant commands.

Finally, to help improve the Army's ability to plan and assess influence operations, we analyzed existing doctrine, organizations, and processes for planning and assessing influence operations, and mapped the metrics-based planning and assessment process we developed in our FY 2005 CAC study into current organizational structures and processes.

Organization of This Monograph

This main portion of this document is organized as follows:

- Chapter Two summarizes the results of our review of commanders' information needs.
- Chapter Three presents a framework for thinking about the sources of commanders' information needs for influence operations and for organizing this information.
- Chapter Four identifies a number of challenges identified as needing consideration by the U.S. Army.

A number of appendixes provide additional information:

- Appendix A lists the information requirements for influence operations that were identified in our structured conversations with commanders and other sources.
- Appendix B identifies key IO-related and influence operations–related tasks and effects, and provides our detailed analysis of influence operations–related tasks in the Universal Joint Task List (UJTL) and Army Universal Task List (AUTL).

- Appendix C sets out a step-by-step process for implementing the study team's metrics-based planning and assessment approach for influence operations.
- Appendix D provides our assessment of the expected utility modeling forecast approach's adaptability for influence operations.
- Appendix E provides our assessment of the SNA approach's adaptability for influence operations.
- Appendix F lists the questions we used for our structured conversations with six senior commanders and one analyst in Washington, D.C., and for more than 30 junior commanders at the Army War College in Carlisle Barracks, Pennsylvania.

The study commenced in the fall of 2005, and project staff briefed the sponsor on the results reported here in September 2006. This report includes information available to the study up until late September 2006, the time at which the project concluded.

CHAPTER TWO

Commanders' Information Needs for Influence Operations

In this chapter, we report the key findings from our efforts to understand the information needs of commanders for information and influence operations in stability operations. We pursued a number of interlocking lines of inquiry, including interviews with commanders, a review of commanders' own writings on IO and influence operations, a case study analysis of recent U.S. operational experience with influence operations, an examination of information needs identified by brigade commanders and their staffs during recent rotations at the NTC, a review of requests for information to the 1st IOC, observations from the UQ 06 exercise, and a review of doctrinal and related publications. (For a detailed explanation of the task list analysis and what it revealed, see Appendix B.)

Insights from Structured Conversations with Commanders

One of the primary sources of data for this study was a set of structured conversations with commanders who executed influence operations in the field in the Balkans, Afghanistan, and/or Iraq. (Appendix A lists the information requirements for influence operations that were identified in our structured conversations with commanders; Appendix F provides the protocol we used for these structured conversations.)

We spoke with a mix of battalion-, brigade-, and corps-level commanders, including both maneuver and support unit commanders. The interviews were conducted during the summer of 2006 in Washington, D.C., and at the Army War College in Carlisle, Pennsylvania. Each interview was done with the aid of a protocol (see Appendix F) comprising five loose themes/discussion areas on which project team members based their questions. Our sample of about 30 interlocutors in all, while neither scientifically selected nor perfectly balanced, provided us with what we think is a representative range of perspectives on commander's information needs for influence operations in today's Army.

Virtually all of the officers we spoke with agreed that their most critical information needs for influence operations pertained to key leaders in the local population and the "pulse," attitudes, or mood of the local population. Understanding the current feelings of local leaders and populations in stability and support operations (SASO) types of contingencies was deemed more important than understanding the enemy's IO/influence operations order of battle or resources, or the technical intricacies of the local information infrastructure (the number of cell towers, radio broadcast footprints, etc.).

One of the more interesting differences of opinion that we discerned among commanders was on the issue of whether information requirements for influence operations should include the monitoring of enemy, or Red, IO themes and messages at all.

We found that battalion- and brigade-level commanders tended to believe that an understanding of Red's IO themes and delivery platforms was relatively unimportant at their level. The consensus of this group was clearly that experience had taught them that if they were able to mount an effective, honest, and truthful influence operations campaign of their own, one that targeted the right local audiences, then they need not worry about what the insurgents' IO effort was producing. Indeed, some of these officers said they thought paying attention to and trying to counter Red IO in any detail would have taken them "off message" and weakened the force of their own influence operations effort.

At the higher echelons, there was much more interest in monitoring and countering Red IO. The commanders at corps level and above with whom we spoke all said they had observed and studied Red's IO messages with interest and had devoted significant energy to working to counter them in some fashion. In one case, a commander told us that he had attempted to use Islamic theology to counter some of the IO messages put out by radical *takfiri* elements in Iraq; this effort enlisted several moderate Iraqi imams (prayer leaders) to attack the Islamic legitimacy of some of the *takfiris'* tactics.[1]

By comparison, the majority of the lower-echelon commanders we met told us that they devoted much time and energy to monitoring and influencing the perceptions of a select number of local power brokers in the belief that the populace in the conservative Muslim societies in which they were working would invariably trust and follow the lead of local elites rather than develop their own views on American forces and their intentions in Iraq/Afghanistan. The types of power brokers monitored and cultivated included imams, tribal sheikhs, municipal government officials, and university/school officials. A minority of the commanders opposed this approach and instead favored direct engagement with the local populace at the grassroots level. The members of this minority tried to use "man on the street" surveys, general behavioral indicators (such as the amount of anti-U.S. graffiti observed on patrol routes), and trends in the number of intelligence tips received from local civilians to gauge whether their grassroots efforts were bearing fruit.

Those commanders who focused on cultivating local elites found that extensive face-to-face meetings were their best tool for IO/influence operations. Face-to-face efforts almost always took a long time to pay dividends (all commanders told us that it took several months to gain the trust of the local power brokers), but the sense was that, in the end, they had a positive effect on the level of violence in the area of operation (AO). Those commanders who favored grassroots work

---

[1] *Takfir* is the mechanism by which some Islamist extremists excommunicate other Muslims, thereby dehumanizing them and making them legitimate targets of violence. We should note that several other officers were very ambivalent about using Islamic theology in U.S. IO efforts; these commanders believed that Islamic theological messages coming from U.S. forces would have no credibility with the Iraqi populace, no matter how well crafted they were.

tended to use civil-military operations (CMO) extensively to provide non-lethal kinetic activities in support of their influence and information operations (repairing schools, handing out candy and soccer balls, digging wells, etc.). In some cases, they also used leaflets and handbills to disseminate key messages and themes.

At the higher echelons, two commanders told us that they monitored popular feelings and attitudes by dividing the population into macro-demographic categories, such as urban Sunnis, rural Shiites, urban Kurds, and conducting polling or attitude surveys among these groups. One lower-echelon commander told us his best tool for gauging the popular mood in his city was a network of informants established by his human intelligence (HUMINT) team. He found that the informants provided him with his only genuine insights into the real concerns of ordinary citizens, and he adjusted his influence operations themes often to respond to these concerns.

Most commanders recognized the important role that international media—especially international Arab media, which are expected to have higher credibility than Western media—could play in influencing the attitudes of the local populace. This was deemed especially important in Iraq. Commanders at the echelons above division were deeply concerned about the ability of international media to affect the conduct of their combat operations by raising international attention and concern to heights that would prevent the United States from continuing kinetic operations at the desired level. The classic example cited of an IO threshold being crossed was the first Battle for Fallujah, in April 2004, during which wildly exaggerated reports of civilian casualties were broadcast by international Arab media, causing such an uproar in the Arab world that political pressure for cessation of the operation grew even before the U.S. Marines were able to complete their capture of the city.[2]

---

[2]   Also, bloody footage of the battle reportedly created tensions with the British, led the White House to closely monitor the situation, and ultimately impelled renewed diplomatic efforts for a negotiated solution. See Jim Krane, "U.S. Steps Back from the Brink in Fallujah," *Associated Press*, May 1, 2004; David Cracknall, "British Fears on U.S. Tactics Are Leaked," *Sunday Times of London*, May 23, 2004; "British Memo Says Heavy-Handed U.S. Tactics Have Fuelled Opposition in Fallujah, Najaf," *Associated Press*, May 25, 2004.

However, only a few commanders actually devoted significant time and effort to cultivating the international Arab press through regularly scheduled meetings with reporters and the distribution to junior officers of set talking points for interviews. These were commanders who determined that the civilians in their sector were watching the international Arab media regularly and being influenced by it. This small number had as one of their main information requirements a constant and ongoing review of the international Arab media coverage from their sector.[3] There was consensus across commanders that engaging those media outlets perceived to be hostile to the U.S. (Al Jazeera, for example) through ongoing dialogue was a far better influence operations strategy than trying to isolate them or evict them from the theater.

We found that commanders commonly followed the tenor of Friday sermons in local mosques as an indicator. Most officers we interacted with considered this to be a very good measure of the local information environment, and many adjusted the content of their influence efforts in response to changes in the tenor of sermons from week to week.

Several commanders also told us that they frequently touched base with representatives of major nongovernmental organizations (NGOs) operating in their AO in order to get honest feedback on the current attitude of the local populace. A few commanders, mostly at the higher echelons, said they specifically monitored the level of anti-American content in local newspapers and TV broadcasts. Lower-echelon commanders appeared to be less interested in these indicators than in the attitudes of local power brokers, metrics of popular behavior, sermons in mosques, and the state of the local infrastructure.

At least one commander noted that local attitudes on American treatment of detainees was something that his unit monitored closely. Perceptions that U.S. forces abused detainees were seen to be very dan-

---

[3]   It should be noted that those officers with the greatest interest in following international media were the ones commanding forces in urban sectors with a significant middle-class population, i.e., in areas with better-educated populations that would be expected to have more interest in international news reporting and to be better able to afford satellite dishes.

gerous to American credibility in that particular sector. As a result, one officer noted that his unit made it a practice to provide all released detainees with food and water, as well as transportation home.

Almost every officer with whom we spoke declared that a level of cultural knowledge and sensitivity was necessary to understand the context of the local information environment and prevent counterproductive messages or themes from being used in influence operations efforts. The specifics of what exactly was meant by "cultural knowledge" was seldom detailed by our interlocutors, however.[4]

Surprisingly enough, we found that many commanders thought that influence operations also had value for the morale of their own soldiers; indeed, at least two commanders with Iraq experience told us that they saw boosting U.S. and coalition morale as a primary function of their influence operations efforts. These officers told us that the constant drumbeat of American TV networks' negative news on Iraq (most of which can be viewed on U.S. bases in Iraq) eroded the morale of their soldiers enough that they felt compelled to use their influence operations talking points on U.S. achievements in theater during their weekly meetings with soldiers in subordinate units. To these commanders, their own troops' morale and those troops' understanding of U.S. goals and their role in accomplishing them were major parts of the local information environment.

Finally, we discovered from the interviews (just as we did from the NTC data) that official unit Commander's Critical Information Requirements (CCIRs) often purposely do not include indicators on the local information environment. Many officers stated that they viewed CCIRs as being "alarms" that would cause the unit staff to wake up the commander in the middle of the night. Most information environment indicators did not fall into this category in the view of these officers. They thought that indicators of the information environment belonged in the category of simple information requirements, which represent the basic "pipeline" mode of commander-staff interaction (discussed in greater depth in Chapter Three).

---

[4]   The subject of training and education for soldiers to improve their "cultural intelligence" was not part of our study.

Insights from Recent Papers by Senior Commanders

Although the influence operations literature is expanding rapidly as the Army increasingly recognizes the importance of non-kinetic activities in stability operations, four papers by commanders recently or currently serving in Iraq have received particular notice. Because of these commanders' commitment to integrating IO and civil affairs operations with combat operations, their thoughts are relevant to identifying the most critical types of information needed to conduct successful influence operations at the operational and tactical levels, and for developing "best practices" and new doctrine and TTP.

GEN Peter W. Chiarelli, Commander, 1st Cavalry Division

In a recent paper, GEN Peter W. Chiarelli and co-author MAJ Patrick R. Michaelis describe the 1st Cavalry Division's integration of IO and influence operations with other lines of operation (LOOs) in a largely Shiite neighborhood of Baghdad called Sadr City in 2004–2005.[5]

Chiarelli describes his division's operational campaign plan as being balanced across five major LOOs, with each LOO tied to IO, effectively constituting a sixth LOO. Two LOOs were traditional military activities: combat operations and training and employment of security forces. The three other LOOs supported nation building, including provision of essential services, promotion of governance, and development of economic pluralism.

Chiarelli and his co-author identify the types of information the division required to support a robust influence operations capability. These include data on the ethnic, religious, and cultural makeup and beliefs of the populace; on key Iraqi stakeholders (e.g., tribal or clan leaders) and on individuals who could facilitate meetings with these stakeholders; on the status of infrastructure and basic services; on economic progress (such as prices, wages, unemployment figures, business activity, and waiting times at gas pumps); on activities of NGOs; on such enemy activities as attack locations, rates, trends, and recruitment;

---

[5]   Peter W. Chiarelli and Patrick R. Michaelis, "Winning the Peace: The Requirement for Full-Spectrum Operations," *Military Review*, July–August 2005, pp. 4–17.

and on media activities. Reckoning that conditions on the ground affected support or opposition for insurgent activity, Chiarelli also requested that correlations between enemy attacks and other activities, on one hand, and local conditions, on the other (for example, substandard conditions or successful infrastructure projects), be monitored.[6]

One important point should be noted with regard to the experiences described in this paper. The infrastructure restoration effort conducted by Chiarelli's 1st Cavalry Division was limited to the Shiite Sadr City district of Baghdad, so one must be cautious about applying the influence operations lessons learned here to other parts of Iraq, especially rural Sunni areas. That said, Chiarelli's efforts provide a useful contribution on how commanders think about measures of effectiveness (MOEs) in stability operations in that he was measuring U.S. activities, conditions on the ground, and the impacts of both on enemy activity.[7]

MG David H. Petraeus, Commander, 101st Airborne Division (Air Assault)

In then-MG (now GEN) David H. Petraeus's view, COIN campaigns are more than just military operations. Recounting lessons learned from his time commanding the 101st Airborne Division, the Multi-National Security Transition Command–Iraq, and NATO Training Mission–Iraq, Petraeus offers 14 observations on how to conduct a COIN operation:

---

[6]   Chiarelli and Michaelis (2005, p. 8) report that "72 percent of the local populace stated there was a direct correlation between their sense of security and the presence of the IPS [Iraqi Police Service]." They also report (pp. 9–12) direct correlations between enemy action and lack of basic services; between level of local infrastructure status, unemployment figures, and attacks on U.S. soldiers; and between terrorist incidents and funding levels in Sadr City.

[7]   JP 1-02 (DoD, 2001a, p. 337) defines *measure of effectiveness* as "[a] criterion used to assess changes in system behavior, capability, or operational environment that is tied to measuring the attainment of an end state, achievement of an objective, or creation of an effect. Also called MOE."

1. "Do not try to do too much with your own hands";[8]
2. Act quickly, because every Army of liberation has a half-life;
3. Money is ammunition;
4. Increasing the number of stakeholders is critical to success;
5. Analyze "costs and benefits" before each operation;
6. Intelligence is the key to success;
7. Everyone must do nation-building;
8. Help build institutions, not just units;
9. Cultural awareness is a force multiplier;
10. Success in a counterinsurgency requires more than just military operations;
11. Ultimate success depends on local leaders;
12. Remember the strategic corporals and strategic lieutenants;
13. There is no substitute for flexible, adaptable leaders;
14. A leader's most important task is to set the right tone.[9]

COIN campaigns thus involve nation building, where "money is ammunition," "cultural awareness is the force multiplier," and success depends on the success of efforts to gain support from local leaders. Indeed, as should be clear from the list, most of Petraeus's observations are directly relevant to the enterprise of influence operations, whether dealing with indigenous leaders, cultural awareness, nation or institution building, or other activities.

According to Petraeus, the key types of information required to conduct such operations include knowledge of local leaders, knowledge of the geographic and the cultural terrain (e.g., ethnic groups, tribes, religious elements, political parties, government structures and processes, local and regional history), and human intelligence.

---

[8] As his first observation, Petraeus is quoting the 15th of T. E. Lawrence's 27 articles, found in T. E. Lawrence, "The 27 Articles of T. E. Lawrence," *The Arab Bulletin*, August 20, 1917.

[9] David H. Petraeus, "Learning Counterinsurgency: Observations from Soldiering in Iraq," *Military Review*, January–February 2006, pp. 2–12.

LTG Thomas F. Metz, Commander, III Corps, Coalition Joint Task Force–7, and Multi-National Corps–Iraq

In this article, LTG Thomas F. Metz recounts his experience with IO and influence operations over a number of years, including his most recent tours in Iraq.[10]

Metz and his co-authors describe the two Fallujah operations to illustrate the power of properly integrating influence operations into the battle plan. In their analysis, Operation Vigilant Resolve (the April 2004 operation) failed because of lack of support from the Interim Iraqi government and heavy international media coverage of unsubstantiated enemy reports of collateral damage and excessive force. By comparison, for Operation Al Fajr (the November 2004 operation), courses of action were developed to mass effects in the information domain and prevent a recurrence of the earlier outcome.

This paper details key influence operations activities and kinds of information required during his command, including: understanding the local populations and key leaders; understanding enemy information centers, networks, and infrastructure; understanding and controlling what Metz and his co-authors call the "IO threshold"—the point at which enemy IO can undermine the coalition's ability to conduct combat operations by creating perceptions that U.S. combat operations are indiscriminate and need to be reigned in; and knowledge of local media and other information channels and of local attitudes, beliefs, and media consumption.

COL Ralph O. Baker, Commander, 2nd Brigade Combat Team, 1st Armored Division

COL Ralph O. Baker, former commander of the 1st Armored Division's 2nd Brigade Combat Team (BCT), provides a recent view of influence operations from the perspective of a brigade commander.

Baker's paper accents the critical role of trust in influence: He devoted considerable effort to winning and maintaining the trust

---

[10] Thomas F. Metz, Mark W. Garrett, James E. Hutton, and Timothy W. Bush, "Massing Effects in the Information Domain: A Case Study in Aggressive Information Operations," *Military Review*, May–June 2006, pp. 2–12.

of local elites and establishing a reputation for telling the truth, and worked through local elites and media because they were most likely to be trusted by the larger population.

According to Baker, careful target audience analysis, the cultivation of Arab media sources, and the collection of information to support measures of effectiveness were crucial strengths of the 2nd BCT's IO efforts.[11] Baker identified five key target audiences based on types—sheikhs, political leaders, academics, etc.—rather than ethnicity, and assessed their attitudes, beliefs, and media consumption habits.

To better understand the diverse ethnic, cultural, economic, religious, and educational dynamics in its AO, the brigade collected detailed demographic information for neighborhoods in the AO. And to identify key leaders and influential groups, it collected information on political and administrative boundaries, organizations, and networks.

The brigade also conducted media content analyses and surveys to better identify the most-popular newspapers and TV stations. Baker hired Iraqis to assist with the information collection process and had them monitor Arab media 24 hours a day, seven days a week to track what was being said about coalition forces. The brigade also periodically monitored enemy IO efforts and messages in order to better counter them. Extensive performance metrics were developed, including favorable/unfavorable reports running on major Arab satellite networks and in major papers; intelligence tips received (more tips indicated a more cooperative populace); anti-coalition/radical content at Mosque sermons; statistics on brigade influence operations activities (how many meetings with leaders, press events, etc., each week); the status of local and national renovation and reconstruction projects; and intelligence on insurgent/terrorist activities (including casualties, property damage, and disruptions to electricity, water, fuel in order to inform the public of their negative impact).

---

[11]   Ralph O. Baker, "The Decisive Weapon: A Brigade Combat Team Commander's Perspective on Information Operations," *Military Review*, May–June 2006, pp. 13–32.

Key Common Insights from Commanders' Papers

Although some commanders emphasized different aspects of influence operations (e.g., Chiarelli focused extensively on infrastructure development while Baker focused more, although not exclusively, on influencing the population by developing ongoing relationships with local leaders) in their papers, all of them demonstrated a keen understanding of and commitment to the importance of integrating IO, influence operations, and other non-kinetic activities into combat operations. All recognized the tremendous challenge of conducting influence operations efforts in areas where the enemy's understanding of the local information environment is significantly better than your own, underscoring the urgency of improving coalition IO.

Another important commonality among the commanders was their emphasis on the need to adequately understand target audiences, including their culture, social norms, and psychology. Without such understanding, it is difficult to craft coalition messages that resonate with the local populace and to convince them that they and the U.S. forces have a mutual interest in stability and reconstruction. In the absence of cultural awareness, it is also difficult to counter enemy IO. A large part of this effort involves ongoing, often consuming, interaction with key local leaders and constituencies. Developing good relationships with local leaders through face-to-face meetings can also assist in putting an Iraqi face on influence operations efforts—another key lesson coming out of these commanders' papers.

Commander involvement and interest in activities related to influence operations was another important aspect of the more successful IO and influence operations efforts, as was communicating to subordinates the importance of conducting IO-related activities on a continuous basis. And finally, monitoring and ensuring the success of key infrastructure projects important to the local population played a significant role in making the population more receptive to coalition messages and more cooperative in countering enemy activity.

Insights from Case Study Analyses

We now shift our focus from a review of four papers by commanders with recent operational experience integrating influence operations into larger combined arms operations to our larger case study analysis of commanders' information needs for influence operations in Bosnia, Kosovo, Afghanistan, and Iraq.

Bosnia

Our review of the literature on Bosnia revealed that this campaign included both a general PSYOP effort and a targeted IO/influence operations effort. The PSYOP effort seems to have been fairly ineffective; the targeted IO and influence operations effort, however, appears to have realized some successes in pushing the process of political reform and democratization forward.

For the PSYOP contribution to influence operations, target audience analysis was not terribly sophisticated, as PSYOP units lacked adequate language skills and regional expertise. As one brigade commander complained, "PSYOP messages were bland, ineffective, and not properly targeted to the local population."[12] In Bosnia, U.S. and NATO PSYOP units employed a wide range of delivery tools, including magazines, newspapers, handbills, and radio and TV stations. All of this notwithstanding, PSYOP had difficulty identifying pivotal demographic or other groups, so theme selection was not appropriately tailored to specific audiences. This resulted in products being generated that often were not culturally appropriate, although there was improvement in this area after U.S./NATO forces had been in country for a while. In short, the written record suggests that PSYOP doctrine did not prove useful in executing PSYOP at the operational level.

In contrast to the general PSYOP effort, the targeted influence operations effort (which included some PSYOP tools) conducted by U.S. forces was, especially after mid-1996, well coordinated and focused, and had positive effects on the stabilization of Bosnia. This

---

[12] See Stephen C. Larsen, "Conducting Psychological Operations in Sophisticated Media Environments," master's thesis, U.S. Army Command and General Staff College, Fort Leavenworth, Kan., 1999, p. 13.

effort included a carefully calibrated mix of radio broadcasts, hand-bill distributions, press conferences, and face-to-face meetings between local elites and senior U.S. commanders that was mapped out to have maximum effect on such key events as national elections and refugee resettlement programs. Target synchronization matrices were employed to aid the timing and calibration of activities.

Planning documents that our project team obtained from an officer involved in the American Bosnia IO effort reveal that after mid-1996, coalition forces tracked a fairly exhaustive set of information requirements on the national information environment. The following is a truncated list of these requirements:

- culture and social structure within ethnic communities (moral codes, norms of community participation, roles and status of family member, lines of authority, aspects of etiquette)
- administration of justice (judicial procedures; expectations for prosecutors, judges, etc.)
- political parties (biographies and personalities of leaders, internal dynamics, relation to current government)
- the armed forces (levels of foreign influence, personalities of key officers, sources of recruitment)
- public education (philosophy guiding the system, requirements for students, teacher education, political influences on the system)
- property rights (nature of property laws and codes, methods for ownership transfer)
- radio and TV networks (numbers and types of transmitting stations, censorship practices, level of propaganda usage, levels of foreign influence, programming and content)
- public works and utilities (condition of public buildings, roads, and housing; condition of power, water, and sewage systems).

In fact, stabilization forces planners achieved a high level of target audience and message differentiation in Bosnia in the 1996–1997 period.[13] They emphasized a total of 10 different themes: six categories of themes related to distinct target audiences, and four categories related to different activities or LOOs.

The six themes related to target audiences were

1. general target audience themes
2. Brcko (city in northern Bosnia-Herzegovina) target audiences
3. police target audiences
4. displaced person/refugee returns target audiences
5. Doboj (city in northern Bosnia-Herzegovina) target audiences
6. "Sapa Thumb" target audiences.

The four thematic categories related to different activities or LOOs were

1. municipal elections themes
2. economic development themes
3. force protection themes
4. equip and train themes.

Examples of messages that included themes ran from the general, such as, "Those who violate the Dayton agreement threaten peace and stability," to the more specific, such as, "You are responsible for controlling your citizens and keeping the peace," a theme directed at public officials in Doboj.[14]

Kosovo

In Kosovo, brigade leaders demonstrated keen interest in and were proactive in using influence operations to further the mission—as evidenced by the commander hosting weekly meetings with Serb and Kosovo Liberation Army leaders to issue policy guidance and promote the overarching theme of restoring peace and normalcy.

---

[13] The source of the material that follows is SYTEX, Inc., *Introduction to Information Campaign Planning and Execution*, student materials handbook produced for U.S. Army Land Information Warfare Activity, Vienna, Va., 1997.

[14] An exhaustive list of specific messages is in SYTEX, 1997.

The IO staff participated in intelligence preparation of the battlefield (IPB), analyzing the use and flow of information to social, civil, political, media, and paramilitary organizations and key leaders. IO staff also identified conduits for engaging target audiences and studied how Albanians and Serbs collected, disseminated, and used information, and how to prevent adverse use of the information infrastructure.

MOEs were based on trend analysis from unit intelligence summaries and operational reports, such as the reporting of negative incidents (e.g., anti–Kosovo Force propaganda, interethnic violence) versus positive incidents (e.g., interethnic cooperation, observance of laws). Media reports were also assessed as positive, neutral, or negative, and themes disseminated by Serb and Albanian media were monitored. Feedback obtained from face-to-face interactions was very important for assessment efforts.

Afghanistan

Most formal brigade CCIRs that we found in the secondary literature were conventional in nature (enemy attacks on maneuver units, enemy attacks on forward operating bases, deaths of soldiers, aircraft crashes, etc.). Influence operations were generally assigned to a unit fire support officer who had little or no previous IO background, leading to an inevitable focus on kinetic issues and solutions. The literature suggests that PYSOP or civil affairs personnel ordered to handle influence operations tasks and themes were often marginalized and not well integrated.

Unit commanders appear to have focused on two macro types of information during SASO in rural Afghanistan. Face-to-face engagements with village elders and religious leaders allowed these commanders to gauge the mood of the local population and its attitudes toward both coalition forces and insurgents. This information was supplemented by more-casual contacts with locals during routine patrolling. All patrols carried cards with influence operations themes in their pockets. A second type of information was more objective and involved the assessment of local infrastructure needs (wells, schools, clinics, etc.) and the most cost-effective options for improving that infrastructure in a way that would provide immediate benefit to the local populace.

In terms of assessment of influence operations efforts, the documents we reviewed suggest that MOEs for stability operations in Afghanistan are still not well developed. Most metrics used are based on inputs or outputs (e.g., number of schools built, number of officials elected, number of meetings held with village elders, amount of reconstruction funds disbursed) rather than outcomes (e.g., effects on popular beliefs, attitudes, and behaviors).

Iraq

As demonstrated by our review of commanders' writings on the subject (discussed earlier), U.S. forces have made some progress in integrating influence operations into operations in Iraq, particularly at the tactical level; but commanders' information requirements are still not being adequately met. Coalition forces are struggling to keep up with their adversaries in understanding and exploiting the Iraqi information environment.[15] Although command interest and influence have been identified as critical enablers for the successful integration of influence activities,[16] not all high-level commanders have demonstrated the requisite interest and commitment to influence operations by specifying

---

[15] The Army appears to be cognizant of this problem, as demonstrated by the issuance of an IO handbook for Iraq based on two lessons-learned reports that highlight the shortcomings of recent efforts and suggest lessons learned and emerging best practices. See Center for Army Lessons Learned (CALL), *Tactical Commander's Handbook, Information Operations: Operation Iraqi Freedom (OIF)*, Combined Arms Center (CAC), Fort Leavenworth, Kan., May 2005b (not available to the public). The two reports on which IO lessons for Iraq were based are both CALL documents: *Initial Impressions Report, Operation Iraqi Freedom: Information Operations, Civil Military Operations, Engineer, Combat Service Support*, Report No. 04-13, May 2004; and *Initial Impressions Report, Information Operations: Information Operations, Organization and Pre-Employment Preparations for Information Operations, Integration of Information Operations Into Planning and Operating*, May 2005a. Both of these are unavailable to the public, but the second is summarized in "Integration of Information Operations into Planning and Operations, Public Affairs, and the Media; Extract from Center for Army Lessons Learned Initial Impressions Report 05-3, Information Operations," Chapter Seven in *Media Is the Battlefield*, CALL Newsletter No. 07-04, October 2006, p. 51. The summary provides overview observations and lessons learned for brigade-level integration of IO, media analysis, audience analysis, the media environment, media engagement, and IO-related doctrine, organization, training, leadership, materiel, and personnel.

[16] See, for example, CALL, 2006. Commander bias toward kinetic operations also has been identified as an issue. See Christopher J. Lamb, *Review of Psychological Operations: Lessons*

influence-related PIRs, and many commanders who have, particularly at the brigade and battalion levels, have not received adequate information support. Because there have been no centralized databases for division commanders and below for tactical influence operations purposes that can be passed from one rotation to the next, commanders in Iraq have had to formulate a picture of the local information environment at the beginning of their rotation and update it by trial and error. For example, the 82nd Airborne Division's "lessons learned" report suggests that "specific information on a specific area (i.e., detailed information on population demographics and city assessments) [was] not available" to the division.[17] Also notable has been the lack of adequate human factors analysis at the tactical level available to commanders, in part because human factors analysis generally has been the province of the Intelligence Community,[18] but also because most IO efforts above division level in Iraq failed to disaggregate target audiences, attitudes, and messaging.[19] And to the extent that human factors analysis has been conducted at the tactical level, our structured conversations suggest that these efforts are somewhat ad hoc and decentralized.

Some successful examples have emerged at the tactical level despite these shortcomings, but there is still insufficient continuity in influence

---

Learned from Recent Operational Experience, Washington, D.C.: National Defense University Press, September 2005.

[17] Cited in Peter A. Sicoli, *Filling the Information Void: Adapting the Information Operation (IO) Message in Post-Hostility Iraq*, School of Advanced Military Studies, U.S. Army Command and General Staff College, Fort Leavenworth, Kan., May 2005, p. 34.

[18] DoD's *Information Operations Roadmap* (October 30, 2003) calls for stronger analytic support from the Defense Intelligence Agency on human factors issues, but the Central Intelligence Agency (CIA) also has reportedly been active in this area. For example, while at the CIA, Dr. Jerrold Post founded and directed the Center for the Analysis of Personality and Political Behavior, an interdisciplinary behavioral science unit that provided assessment of foreign leadership and decisionmaking for the President and other senior officials to prepare for summit meetings and other high-level negotiations, and for use in crisis situations. See Jerrold M. Post's resume (Post, undated).

[19] According to one source (Baker, 2006, p. 16), "IO planners at commands above division level appeared to look at the Iraqis as a single, homogeneous population that would be receptive to centrally developed, all-purpose, general themes and messages directed at Iraqis as a group."

operations best practices from one commander to the next. As noted above, COL Baker executed a comprehensive influence operations plan in his AO and managed to acquire critical information through ongoing, personal engagement with local leaders. Moreover, current PSYOP doctrine dictates that PSYOP units formulate detailed priority intelligence requirements (PIRs) to focus the collection of relevant information, and the intelligence annex for PSYOP planning is supposed to include such factors as enemy disposition, anticipated opponent PSYOP and information plan, population status, media infrastructure, language analysis, religion analysis, ethnic group analysis, weather analysis, terrain impact on dissemination, a reconnaissance and surveillance plan, and an area study.[20] Thus, efforts to improve commanders' information for the conduct of influence operations already are apparent, but they need further reinforcement and codification in doctrine, education, and training.

In terms of commanders' information needs for assessment of influence operations, there has been some improvement in developing metrics to aid in targeting and to measure effects rather than outputs. For example, in his article, COL Baker identified media analysis, including surveys to identify popular newspapers and TV stations, as important sources of data for guiding influence operations. And GEN Chiarelli, who is widely known for his efforts to demonstrate the impact of CMO, and who developed data showing a correlation between the number of infrastructure projects and job programs and the number of insurgent attacks in his AOR, identified a wide range of indicators of importance to commanders, including ethnic, religious, and cultural factors; key Iraqi facilitators and stakeholders; the status of infrastructure and basic services; information on economic activity; activities of NGOs; and data on enemy activities (such as attack locations, rates, trends, recruitment, and propaganda activities). Nevertheless, a great deal of work remains to ensure that deploying units are organized, trained, and equipped to identify, collect, and assess

---

[20] See Headquarters, Department of the Army, *Psychological Operations,* FM 3-05.30 (MCRP 3-40.6), Washington, D.C., April 2005d, pp. 5–21, and *Psychological Operations Leaders Planning Guide,* GTA-33-01-001, November 2005e.

influence operations–related assessment data. (The issue of metrics for influence operations is taken up in Appendix C.)

Key Trends and Contrasts

Our overview of commanders' information needs in four cases and an overview based largely on secondary sources point to several fundamental, overlapping themes. First, inadequate information on the local population and culture—i.e., the lack of a proper analysis of the target audience—proved to be one of the greatest obstacles to effective influence operations efforts. This lack of sufficient information led to bland messages that did not resonate with specific target audiences and made it difficult to compete with adversaries more capable of exploiting the local information environment.

Another key theme is the crucial role of the commander in setting the tone for integrating influence operations into combat operations, at all echelons, including at the strategic and operational levels. But IO/influence operations understanding and commitment are becoming increasingly critical at the division level and below (particularly among brigade and battalion commanders) because of the tendency of influence operations efforts to be conducted at the unit level. In fact, our structured conversations and other research suggest that commanders at the tactical level typically think that they have the clearest sense of what messages will resonate with their target audiences, that message guidance from division and higher echelons frequently fails to resonate with local audiences, and that approvals of products frequently are too slow in coming to be useful. The result is that brigades and battalions are increasingly developing a new set of best practices in conducting influence operations on their own, thereby largely overcoming the constraints imposed by doctrine that would reduce their effectiveness and agility. How best to resolve the implicit tension between the desiderata of a top-down strategic and operational planning process, on the one hand, and responsive and flexible tactical influence efforts, on the other, is likely to be a recurring question for commanders and their staffs. (Appendix C sets out a step-by-step process for implementing a top-down, metrics-based planning and assessment process.)

Finally, although some progress has been made in the area of MOEs, commanders are still not satisfied that effective assessments of their influence operations efforts are taking place. It is much easier to establish correlations between activities and effects than to establish that specific activities caused those effects.

In addition to the obvious sociopolitical and cultural differences among the cases, there is a difference in the attitudes of the target populations. In Bosnia, Kosovo, and even Afghanistan (at least in the early stages), the local population was considerably more favorable and open minded toward coalition forces and thus more receptive to their messages than is the case in Iraq. Of course, the goodwill of a population can easily dry up in the absence of tangible progress in areas such as reconstruction and security, but coalition forces in Iraq had a smaller margin of error than did the previous cases for losing the target audience, making the influence operations effort there considerably more challenging.

Insights from the National Training Center

We also sought to understand brigade commanders' information needs for IO and influence operations through the lens of the NTC.

The NTC simulates the Iraq experience with Arabic-speaking role players, nine prototypical Arab villages and towns, and sophisticated insurgent forces. Commanders are limited in the IO tools they can employ at the NTC, since rotating units do not have access to radio, TV, or Internet broadcast equipment. Also, the PSYOP teams attached to each brigade are typically very small and have limited reach.

To better understand how brigade commanders viewed their information requirements, we examined the operational order briefings and commander's guidance for five recent NTC brigade rotations. We found that almost all of these documents reflected a great interest at the brigade level in leveraging influence operations to accomplish the COIN mission. Indeed, our discussions with IO trainers at the NTC revealed that as of the summer of 2006, very few brigade commanders who train at the NTC are ambivalent about IO—and the large majority are quite enthusiastic about the potential of influence operations in Iraq and Afghanistan.

Our analysis of PIRs for the five brigade rotations we examined suggests that commanders specified an initial set of PIRs that could provide them with a baseline regarding key features of the operating environment, but then tended to shift to viewing PIRs as events and other information they considered to be urgent indicators.

The first category, indicators that commanders identified to provide a baseline assessment of their operating environment and its information domain, included the following sorts of information:

- structure and leadership of insurgent groups
- Iranian influence in the AO
- enemy propaganda efforts and activities
- nature of the local media infrastructure (e.g., radio transmitters, TV stations, newspapers)
- linkages between local institutions (e.g., mosques, schools) and insurgent recruiters.

Once well situated with a baseline assessment, many commanders and brigade staffs apparently viewed PIRs as urgent indicators, or "alarms," that required immediate commander attention. Thus, as we delved further into the raw data received from the NTC, we found that many of the information requirements sought by rotating units were not actually included in the PIRs listed: Day-to-day data on changes in the information domain generally were not seen as having the requisite urgency in many cases to qualify as PIRs and were generally treated as part of a routine "pipeline" reporting process.[21]

That said, various mission preparation briefings for rotating units revealed a wealth of indicators on the information domain that were identified as being of interest to commanders and their staffs just below the threshold for PIRs. Some of the principal indicators we found were

---

[21] Alarm and pipeline modes are discussed in James P. Kahan, D. Robert Worley, and Cathleen Stasz, *Understanding Commanders' Information Needs*, R-3761-1, Santa Monica, Calif.: RAND Corporation, 2000.

- attitude of a village's mayor, police chief, and imam toward Blue forces
- village's unemployment rate
- village's availability of electricity or potable water
- state of the village's primary school (e.g., is it in any disrepair?)
- attitude of the major international news networks toward Blue.

We found it interesting that the NTC PIRs focused heavily on the IO activities and links within Red, whereas the lower-level indicators (which were discussed in great detail and assessed thoroughly) focused heavily on the attitudes of local power brokers, neutrals, and friendly parties, and on the state of local infrastructure.

Insights from 1st Information Operations Command

The Army's 1st IOC, which is headquartered at Fort Belvoir, Virginia, is the operational Army's main point of contact for advice/guidance on the conduct of influence operations in the field. Currently, units deployed in Iraq and Afghanistan can obtain assistance from the 1st IOC in two ways. They can draw on the expertise of deployed field support teams from 1st IOC that can deploy and work directly with battalion, brigade, division, or corps staffs to develop and critique IO campaign support plans; or they can electronically send reachback requests for information to Fort Belvoir, where the IO analysis support teams of the 1st IOC work to answer these as completely as possible.

During our visit to the 1st IOC, we discussed with analysts and managers various aspects of IO doctrine, as well as influence operations practice in the Balkans, Iraq, and Afghanistan. On the subject of information requirements in the field, we gleaned three insights of particular importance from these discussions.

We learned that most of the reachback requests from the Afghan theater that are filed have to do with "hard" information requirements, i.e., data on the technical characteristics of Afghanistan's information infrastructure. These requests focus on questions about such matters as radio transmission footprints, the size of local TV transmitters, the geographical penetration of TV broadcasts from Iran and Pakistan, and the rate of growth of the cell phone network in a certain province.

Our discussions further revealed that when requests about "soft" information requirements are sent back to the 1st IOC from Afghanistan, they are often appeals for help on rapid-response IO efforts being mounted after major events that grab media attention, such as the kidnapping of a foreign aid worker by the Taliban or a traffic accident involving coalition forces that results in the death of Afghan civilians. Commanders in the field want several courses of action (COAs) for themes/messages they can put out to either protect the reputation of coalition forces or degrade/weaken the local standing of the Taliban and its allies.

Finally, our interlocutors told us that a number of information requests from commanders in Iraq were queries about the kinds of chatter appearing on Europe-based jihadist Web sites about terrorist TTP. Many commanders in Iraq think that there is a steady migration of terrorist TTP from the radical Islamist community in cyberspace (much of which is physically located in Europe) to the battle space in Iraq. While much of this interest in jihadist Internet chatter is concerned with kinetic TTP and innovations, there is undoubtedly a desire to learn about the new concepts for jihadist media operations that are being examined, assessed, and debated on radical Islamist Web sites and in chat rooms. There is a belief that the hypothetical media strategies being debated in cyberspace today could be employed for real in Iraq in a few months' time. Thus, we see that commanders' information requirements in Iraq can indeed extend beyond the geographic confines of Iraq to include a basic understanding of developments in the TTP incubators that many Europe-based radical Islamist Web sites have become.

Observations from Unified Quest 2006

UQ 06 was a major joint war game based on fictional parallel global and regional conflicts set in the year 2015 that sought to incorporate lessons learned from recent experiences in Iraq and Afghanistan.[22] Although war games do not actually generate empirical evidence or proofs, the UQ 06 game generated some insights into current joint thinking about influence operations.[23]

In the early moves of UQ 06, Blue commanders in the Southeast European Federation paid scant attention to the mechanics of both executing IO in the various provinces of the AO and integrating it into the main LOOs. They were aware that influence operations were important and needed in order to increase popular interest in participating in elections, give civilians more reasons to support the pro-Western government, and convince more locals to provide Blue with information on insurgent hideouts and tactics. However, their guidance to tactical units was too general to be of much use, and influence operations did not do a particularly good job of integrating kinetic and politico-military measures along with non-kinetic ones. Key target audiences and their characteristics were not specified; as a result, generic influence operations messages were disseminated to an audience that had not been broken down according to demographic groups or other key factors that can make a difference in persuading different audiences.

Moreover, the messages initially produced were overly broad and not directive in nature—they did not truly encourage the population to take specific actions, nor did they show how specific actions might lead to good outcomes for ordinary citizens (e.g., less violence on the streets, more consumer goods, more influence on local governance). Mechanisms for delivering messages were also not specified to tactical commanders—there was no sense given as to what the mix between leaflets, handbills, face-to-face engagements, media interviews, etc., ought to be in a given sector. Perhaps worst of all, the intelligence requirements laid out for each task force in the Southeast European Federation included no indicators related to the information environ-

---

[22] UQ 06 was cosponsored by TRADOC and the U.S. Joint Forces Command and held at the U.S. Army War College, Carlisle Barracks, Pa., in April 2006. See Gary J. Gilmore, "Afghanistan, Iraq Lessons Learned Part of Joint War Game," *Armed Forces Press Service*, March 28, 2006.

[23] On the possibilities and limits of war gaming, see Herman Kahn and Izi Man, "War Gaming," P-1167, Santa Monica, Calif.: RAND Corporation, 1957.

ment. Thus, there was no mechanism or process through which intelligence officers could tell the Blue leadership whether influence efforts were having any effect. Taken together, the rather superficial treatment of influence operations in the early phases of the game suggested that most game participants initially had great difficulties conceiving of how best to integrate them into combined and joint actions.

In the later moves of UQ 06, Blue improved its influence operations efforts considerably. Target audiences were specified, and messages became directive and specific. Metrics for assessing Blue performance and effectiveness in influence operations even appeared in an embryonic form; and guidance on the use of delivery platforms started to appear. The lesson from this was that influence operations efforts need to be mechanically well planned from the outset of a stability operations contingency.

Although the view of influence operations changed somewhat over the course of the games, the games suggest that much of the joint community continues to view influence operations as a somewhat marginal and separable set of activities typically not planned, executed, or assessed as essential parts of larger joint COIN or SSTR operations, much less global shaping operations. There was, accordingly, little attention to metrics or other information that might be needed to plan, execute, and assess influence operations as parts of larger joint operations.

The UQ 06 exercise led to two principal concerns about influence operations that are germane to our study. First, the role of influence operations—and information needs for influence operations—may currently lack a common frame of reference in the thinking of joint warfighters and may not be terribly well crystallized in the joint community as a whole. Second, we worry that a failure to fully and effectively integrate influence operations into the joint and combined campaign, and inattention on the part of Blue commanders to the mechanics and specifics of influence operations, could be very damaging to a larger COIN or stability operation, where the operation's success hinges on the effectiveness of influence operations.

Insights from a Review of Doctrine, Tactics, Techniques, and Procedures, and Task Lists

No analysis of commanders' information needs for influence operations would be complete without a review of doctrine and TTP and an assessment of the formal tasks associated with these activities.

The two principal documents that govern Army IO are JP 3-13, *Information Operations*, and Field Manual (FM) 3-13, *Information Operations: Doctrine, Tactics, Techniques, and Procedures.*[24]

As pointed out in our discussion of definitions in Chapter One, there is as yet no accepted joint or Army definition of (or Army or joint doctrine for) influence operations, and no clear doctrinal foundations for some of the principal military activities that arguably are part of influence operations—notably, STRATCOMM, but also military diplomacy and defense support to public diplomacy. Nevertheless, joint and/or Army doctrine is well developed for other influence operations-related activities, including CMOs, civil affairs (doctrinally, a subset of CMOs), and public affairs.

An examination of JP 3-13 and FM 3-13 reveals a well-developed body of doctrine for planning, preparing, executing, and assessing IO. Among the many aspects the documents address are intelligence, surveillance, and reconnaissance (ISR) contributions to information superiority, and the interdependence of IO and intelligence functions in profiling the information environment during the campaign planning process.[25] A short section in FM 3-13 calls for building databases comprising the kinds of information needed to support IO "across the spectrum of conflict"—i.e., in peace, crisis, and war. All of this would suggest that the joint and Army IO doctrine publications are cognizant that commanders have particular information needs for IO and

---

[24] DoD, *Information Operations*, JP 3-13, Washington, D.C., February 13, 2006a; and Headquarters, Department of the Army, *Information Operations: Doctrine, Tactics, Techniques, and Procedures*, FM 3-13, Washington, D.C., November 2003c.

[25] FM 3-13 describes the kinds of information needed for IO and states that G-7 submits information requests to G-2 to fulfill those needs; during IPB, G-7 works with G-2 to determine adversary IO capabilities and vulnerabilities (Headquarters, Department of the Army, 2003c, pp. 1-10 to 1-11, 5-6, and 5-9 to 5-12).

have attempted to develop the necessary framework for accommodating those needs. Nevertheless, there is reason to worry that the framework provided in JP 3-13 and FM 3-13 may not adequately support commanders in the kinds of operations the Army is engaged in today, including COIN and stability operations. According to the introduction of FM 3-13, for example:

> Information Operations (IO) encompass attacking adversary command and control (C2) systems (offensive IO) while protecting friendly C2 systems from adversary disruption (defensive IO). Effective IO combines the effects of offensive and defensive IO to produce information superiority at decisive points.[26]

In keeping with this conception, joint and Army IO doctrines designate a nearly identical set of core and supporting elements of IO focused almost exclusively on denying enemy leaders and forces the information needed to make timely and accurate decisions while protecting the quality and timeliness of friendly information and denying the enemy access to it. Such a conception of IO may be appropriate in major combat operations, where making faster decisions based on higher-quality information gives a commander an advantage in conventional warfare; but the ability to gain information superiority over an enemy command and control (C2) system is less relevant in stability, reconstruction, and COIN operations—the very activities in which the Army is so heavily engaged in Afghanistan and Iraq today.

In those kinds of operations, the most important targets of influence are not enemy commanders, but individuals and groups, both local and international, whose cooperation is vital to the mission's success. Granted, joint and Army IO doctrine publications do not ignore these targets—PSYOP and counterpropaganda can be designed to influence them. But it is notable that the activities most directly aimed at influencing local and international audiences—functions such as public affairs, civil affairs, CMOs, and defense support to public diplo-

---

[26] Headquarters, Department of the Army, 2003c, p. v.

macy—are treated only as "related activities" in IO doctrine, if they are mentioned at all.

A number of tasks and effects associated with IO and influence operations are described in two documents that break military activities into standard military tasks: the UJTL and the AUTL.[27]

An analysis of IO-related and influence operations–related tasks contained in the UJTL and AUTL confirms this assessment: Tasks associated with the doctrinal core and supporting elements of IO are described almost exclusively in terms of attacking enemy information systems or protecting friendly information from attack; tasks most relevant to influencing the targets of greatest importance in stability, reconstruction, and COIN operations reside in functions almost entirely outside the doctrinal boundaries of IO.

These findings suggest that current joint and Army IO doctrine is insufficiently broad to support commanders' information needs for influence efforts in settings other than major combat operations, or that it emphasizes the wrong "pillars" for COIN and related operations, where PSYOP and a number of other "supporting" and "related" activities are of greater importance. There are several possible ways to correct this deficiency.

One approach would be to rewrite JP 3-13 and FM 3-13 to address a wider range of operational settings. The information superiority paradigm could be preserved for conventional warfare, with additional chapters added to address the conduct of IO in environments where the principal objective is to influence actors other than enemy leaders and forces. However, such an approach would require broadening the overarching concept of IO to more explicitly address essential links to CMOs and other related activities outside the canonical IO tool set, and it would risk making two already lengthy publications unwieldy. Given the recency of the revision of JP 3-13, it seems highly implausible that it will be rescinded any time in the near future. The Army

---

[27] See DoD, *Universal Joint Task List (UJTL)*, Chairman of the Joint Chiefs of Staff Manual (CJCSM) 3500.04D, Washington D.C., August 1, 2005; and Headquarters, Department of the Army, *The Army Universal Task List*, FM 7-15, Washington D.C., August 2003b. Appendix B provides a list of IO-related and influence operations–related tasks and effects from these documents, as well as a detailed analysis of these tasks.

thus also might consider rewriting FM 3-13 to better capture differences in the emphasis placed on influence operations in different types of operations.

As an alternative, doctrine developers could write additional, separate publications to better define the role of IO and its relationship to influence operations in settings where stability, reconstruction, and COIN operations are the main focus of effort. Such an approach would preserve the information superiority paradigm for conventional warfare while allowing for the development of new doctrine more relevant to the current security environment. It would, however, risk further fragmenting and stovepiping a mission area that some Army commanders complain is already insufficiently integrated with the overall operational effort.

Consequently, the best solution might be to insert more comprehensive IO and/or influence operations chapters in doctrine publications that guide operations in specific settings, such as JP 3-06, *Joint Doctrine for Urban Operations*, and FM 3-07, *Stability Operations*.[28] In this regard, the final version of FM 3-24, *Counterinsurgency*, provides a good model for the explicit treatment of IO in the context of undertaking a larger mission.[29] This approach would enable responsible offices to develop doctrine better tailored for information environments peculiar to specific operational settings while keeping IO and influence operations programs closely integrated with broader operational efforts.

While a thorough analysis of all UJTL and AUTL tasks to determine whether they adequately support commanders' information needs for influence operations was beyond the scope of this study, our summary examination of those tasks suggests to us that they do. Existing tasks appear to address all major influence operations functions, and the associated task descriptions and measures of performance (MOPs) seem to indicate that appropriate interfaces exist to provide command-

---

[28] DoD, *Joint Doctrine for Urban Operations*, JP 3-06, Washington, D.C., September 16, 2002; and Headquarters, Department of the Army, *Stability Operations*, FM 3-07, Washington, D.C., October 2008.

[29] Headquarters, Department of the Army, *Counterinsurgency*, FM 3-24, Washington, D.C., December 2006c. (JP 3-06 and FM 3-07 are both named in this manual.)

ers the relevant information needed to execute those operations effectively. Similarly, a cursory survey of joint and Army doctrinal publications would suggest that well-developed doctrine exists for most key activities related to influence operations. However, by segregating functions related to IO and functions related to other influence operations into separate doctrine manuals—coincident with IO doctrine's overemphasis on the conventional-warfare, information-superiority paradigm—the Army may have allowed different philosophies and organizational cultures to develop among the specialists charged with carrying out those tasks. If so, then the potential exists for different functional specialists to interpret tasks related to IO and influence operations differently, opening seams in the information provided to commanders and potentially resulting in multiple, independent influence efforts that are contradictory and self-defeating. The Army should examine this issue closely and consider how better to integrate IO and influence operations doctrine with the doctrine for operations other than major combat.

Chapter Conclusions

Our review of a range of sources provided us with a number of insights about commanders' information requirements for IO. Here we attempt to distill these insights into a few main conclusions. (For a full, detailed list of the information requirements identified, see Appendix A.)

Perhaps the most important conclusion is that in the types of contingencies in which the U.S. Army now finds itself (COIN and stabilization operations), the most critical information requirements have to do with understanding the attitudes, beliefs, and mood of the local civilian population. Understanding the nature of the technical information and communications infrastructure, studying the messages being put out by the local and international media, and assessing the adversary's IO capabilities and strategies are all helpful, but they cannot substitute for a good understanding of the pulse of the local populace. It should be noted that understanding the popular mood is not and never can be a one-time exercise. Because the popular mood can shift quickly, its key indicators require continuous monitoring—perhaps more so in Muslim societies in which the populace are innately suspicious of the West and the United States and in which rumors from "the street" play such a prevalent role. Shifts in popular opinion are especially likely after a single traumatic incident, whether it is a bombing raid that causes severe collateral damage to civilian homes and property or a traffic accident in which U.S. military vehicles accidentally kill a local child.

Our research shows that success in IO indeed depends on commander interest and involvement and on the commander's image of the battle space and what needs to be done with IO to achieve the desired end state. Commanders who insist that their subordinates conduct a set program of IO activities and who follow up to make sure the program is carried out are far more likely to succeed in integrating IO into the campaign than are commanders who are more passive. Additionally, commanders need to emphasize the importance of IO on a regular basis throughout a unit's yearlong tour of duty in theater.

Our conversations with commanders and some of our other research also revealed that there is no single correct answer to the question of which sources of information ought to be drawn upon in order to accurately assess the local information environment. Instead, these appear to be entirely specific to the mission, the context, and even the commander. Some commanders, for example, have chosen to engage local power brokers to get the pulse of the population, whereas others have preferred grassroots engagement with the local population through patrolling and face-to-face encounters. Other commanders have relied on the behavior of the local civilians from a distance—the amount of graffiti, facial expressions, and hand waves as U.S. troops pass by, for example—as their primary source of data on local attitudes, or have monitored local newspapers and TV stations for pro- or anti-American content. And some commanders have chosen to set up detailed human intelligence networks in their AO and to use the resulting flow of informant reports to gauge the true feelings of the community.

Each of these approaches can offer insights—dependent on the nature and culture of the local area, the resources available to the commander, and the specific military objective at hand—and some combination of these approaches is probably desirable when it is feasible. Also important is that a clear information sourcing strategy be put in place very early on in an AO so that subordinate commanders know what is expected of them over the long term; rapid shifts in this strategy will create confusion in the ranks.

We found that commanders who thought they had employed IO successfully invariably had a clear, uncluttered picture of the key IO and influence operations variables in the current battle space (of which there are typically maybe three or four), a good understanding of the level of resources available to support IO, and a solid vision of what the desired end state of the battle space at the conclusion of the tour of duty was. These officers almost always saw violence and insurgent strength in their AO decline during their year in theater. Although the evidence is anecdotal, commanders who tried to monitor too many variables, who shifted IO resource levels back and forth in response to daily crises without a long-term steady state, or who changed IO themes and messages randomly without any underlying concept of a step-by-step path to victory—these commanders appeared to enjoy less success.

Developing good MOEs to assess how a unit's influence operations are being received by the local population is one of the thorniest problems facing the Army today. Although none of our interlocutors thought that the Army has a particularly good set of MOEs for influence operations in COIN operations and SASO, our interviews revealed that three standard indicators in particular are being used across units and echelons: the tenor of sermons in mosques, the "on the street" behavior of the locals (obscene gestures toward U.S. troops, amount of anti-American graffiti, etc.), and trends (either upward or downward) in the number of intelligence tips from the local population.

Some commanders also noted the highly disruptive impact of unit rotations. Frequently, a new commander will come in and, rather than continuing influence operations programs that by all accounts are working reasonably well, will immediately make major changes to the program or, worse yet, revert to cordon-and-search operations. This can sow confusion and mistrust among the local population and erase any gains that have been made. One commander compared U.S. operations in Iraq with those in Vietnam in this regard: Rather than fighting a multiyear war and accumulating and applying relevant experience over that period, commanders have been fighting a series of six-month actions that generally have failed to identify success factors and to emphasize initiatives that are working.

The findings presented here suggest that commanders increasingly are coming to believe that the success of military operations hinges on the successful integration of IO and influence operations into combined arms actions, and that their success in turn hinges on a systematic and detailed understanding of "soft," or "human," factors—both quantitative and qualitative—that are the province of such diverse disciplines as psychology, social psychology, sociology, political science, communications research, SNA, and economics.

Our interviews and other work indicate that such factors include correct identification of the leaders and groups whose support is essential to success; an accurate understanding of their preexisting goals, constraints, beliefs, attitudes, behaviors, and networks; knowledge of their preferred media and other information sources; and a subtle understanding of societal hierarchies, norms, and negotiating behavior. Unless one uses a definition so elastic that it is essentially meaningless, the term *culture*—which is in the province of the cultural anthropologist—does not quite do justice to the range of factors that must be weighed in planning, executing, and assessing IO and influence operations.

Thus, much more clarity in what is meant by *cultural intelligence, human factors,* and *soft factors* is needed. Most of the writings and officers we dealt with declared that cultural knowledge and intelligence were critical to understanding the information environment in SASO missions, yet very few of them could offer a clear definition of *cultural intelligence* and its elements. Without additional specification, internal Army efforts to develop cultural intelligence are likely to become muddied and unclear.

To this end, in the next chapter we provide a framework that can be used for thinking about commanders' information needs for IO and influence operations—including cultural, human, or soft factors—in a coherent way, and for organizing IPB and combat assessment for these operations.

CHAPTER THREE

Sources of Commanders' Information Needs

Our conversations with commanders and review of the written record suggest that commanders' information needs generally flow from an interaction of factors within three principal arenas: commanders' guidance; the operating environment, including the information domain; and the resources available to the commander. We discuss in this chapter each of these arenas, devoting our attention primarily to the one most relevant to our study: the operating environment, including the information domain.

Commanders' Guidance

The first source of commanders' information needs is the higher-level guidance that commanders receive regarding their overall mission and other matters. Influence operations planning should flow from the top down, designed and executed in support of coherent politico-military objectives and synchronizing non-kinetic and kinetic activities, whether conducted by the services or by other DoD or interagency actors. To ensure relevance and responsiveness, however, tactical-level commanders should in turn have the authority, latitude, and flexibility to adapt the broad planning guidance they receive from higher echelons to meet local conditions. How the tensions between these two desiderata are best balanced will vary from operation to operation.

Our work suggests that effective IO and influence operations require a number of types of higher-echelon guidance that must be provided by superior commanders to subordinate commanders, and that this guidance should include the following:

- a statement of the policy vision, the commander's vision of the end state, and how ends, ways, and means will be brought together in the operation
- specification of political and military objectives
- an expression of the commander's vision and intent
- information on higher-level activities and results
- the commander's specified tasks
- harmonized strategic communications (STRATCOMM), public affairs (PA), and psychological operations (PSYOP) guidance;
- an enumeration of the CCIRs
- an enumeration of the commander's key metrics for assessing campaign-level and tactical outcomes in order to ascertain how well the commander's vision is holding up as the operation unfolds.

The Operating Environment and Information Domain

The second source of commanders' information needs is the commanders' need to understand aspects of the environment—and, especially for information IO and influence operations, of the information environment—that can present opportunities and facilitate mission accomplishment or can introduce challenges that hinder mission accomplishment.

Numerous phrases and initialisms currently in use in the national security community compete for our attention in capturing the most salient features of the contemporary operating environment—especially various "human," "soft," or "cultural" factors—and many of them provide a basis for establishing crude taxonomies of the sorts of information that can (or should) be collected during IPB or in support of ongoing operations.[1] The lack of agreement on a taxonomy of fac-

---

[1]   The phrases include "complex environments," "cultural environment," and "cultural intelligence preparation of the battlefield"; the initialisms include COE (contemporary operating environment), DIME (diplomatic, information, military, economic), METT-TC (mission,

tors that affect influence operations, however, arguably has impeded the ability of military organizations to collect, archive, and analyze the sorts of data needed to support IO and influence operations.

We conceive of the contemporary operating environment as being characterized by three major features of interest: the battlefield environment; the threat; the information domain.

The Battlefield Environment

We include in the battlefield environment the following sorts of factors:

- *geography*: terrain, climate, boundaries, urbanization, and infrastructure
- *human geography*: demographic, social, cultural, and political
- *economy*: type, resources, per capita gross domestic product, employment, industries, land ownership, currencies, and criminal activities
- *links to other regions*: social, cultural, political, and economic.

The Threat Domain

We determined that the following sorts of activities need to be conducted to evaluate the threat:

- Identify opposing individuals, groups, forces, and leaders.
- Assess goals, motivations, capabilities, and vulnerabilities.
- Evaluate enemy sanctuaries, sources of support (financial, moral, logistical), and affinities with targets of influence.
- Determine threat COAs (e.g., identify enemy strategy and assess strengths, vulnerabilities, and possible counterstrategies).

---

enemy, terrain and weather troops available, and civilian considerations), PMESII (political, military, economic, social, infrastructure, information), DPEG (demographic, political, economic, geographical), and ASCOPE (areas, social structure, culture, opportunity, power and authority, economy).

The Information Domain

We define the information domain of the operating environment—which in many important respects captures the major elements of any society that commanders need to be sensitive to in conducting influence operations—as broadly including the following sorts of factors:

- relevant friendly, enemy, and other individuals, organizations, and systems that collect, process, disseminate, handle, or use information, locally, regionally, or globally
- impacts of other elements in the operating environment on the flow, acceptance, or use of information.

As our focus is primarily on IO and influence operations, this is the category of information to which we devoted the most attention.

We now elaborate on the information domain of the operating environment, as well as on how the Army might focus its intelligence and data collection and analysis efforts to support influence operations.

As described earlier, the sorts of data and intelligence most important to commanders in any given operation are context specific and influenced by the mission, commander, and various local factors. Our research suggests that one profitable way of thinking about commanders' information needs for influence operations is to view the problem through a series of complementary lenses that can be used to unpack the most salient features of the operating environment. These lenses can help to identify information that can be organized geospatially, information best organized in terms of networks, and information tied to specific political or military stakeholder groups or their leaders.

Geospatially Oriented Information. Our research suggests that many characteristics of the information domain of the operating environment are best portrayed geospatially, as a set of overlapping layers. Figure 3.1 illustrates this portrayal. As can be seen, one can sensibly place different geospatially oriented features on different planes, or layers. At the foundational (bottom) level, labeled "terrain" in the figure, are physical features of the terrain (such as level of urbaniza-

Figure 3.1
Geospatially Oriented Aspects of the Information Domain

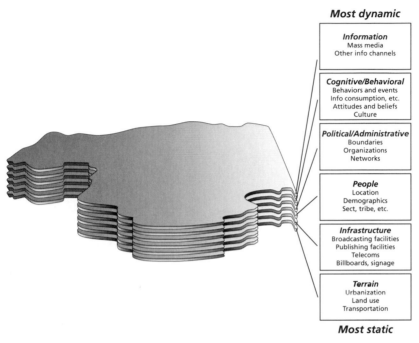

Most dynamic

Information
Mass media
Other info channels

Cognitive/Behavioral
Behaviors and events
Info consumption, etc.
Attitudes and beliefs
Culture

Political/Administrative
Boundaries
Organizations
Networks

People
Location
Demographics
Sect, tribe, etc.

Infrastructure
Broadcasting facilities
Publishing facilities
Telecoms
Billboards, signage

Terrain
Urbanization
Land use
Transportation

Most static

RAND *MG656-3.1*

tion, land use, and transportation networks) that most analysts would acknowledge directly influence the nature of the information faced by a typical population. For example, the physical environment faced by an urban dweller—filled with buildings and streets that accommodate high population densities, and high levels of commercial and other interactions—is simply very different from that faced by a farmer in a remote village. Moreover, economic and other behavioral patterns—as well as a variety of cultural or other characteristics—also may be closely associated with the terrain in which populations live and work. Simply put, the physical layer provides a key foundation for the information domain.

In the second layer of Figure 3.1, labeled "infrastructure," are various features of the infrastructure that are closely related to the information domain, including broadcasting and publishing facilities, telecommunication towers, hubs and switching stations, and various billboards and signage that may clutter the landscape and bombard the individual with a variety of commercial, public service, or other messages.

Because the concentration and disposition of individuals of various races, ethnicities, religions, sects, demographic groupings, tribes, and political and ideological leanings is of critical importance for target audience analysis, people are situated in the third layer (labeled "people").

The fourth layer (labeled "political/administrative"), is an overlay of political, administrative, and other, less formal (e.g., tribal, or commercial) boundaries that regulate interactions between members of different political entities or other groupings, as well as the organizations (e.g., governorates or ministries) that administer various activities within these boundaries, and the networks for those organizations (e.g., regional administrative offices that report to national offices, and the connections between senior bureaucratic or administrative offices or managers).

The fifth layer (labeled "cognitive/behavioral") is closely tied to the groups and individuals ("people") who make up the population in any given region, state, province, or governorate, city, or other geographic area. In this layer, we can relate prevailing behaviors and events (e.g., attacks on civilians or infrastructure), differences in preferred media and other information channels used by residents, and differences in residents' attitudes, beliefs, and cultures.[2]

The final, top layer (labeled "information") is used to capture the local, regional, or national media "footprint," or penetration, whether for satellite or terrestrial TV or radio, for Internet access, or for newspapers, magazines, and other print media. Of interest in this layer is the prevalence and penetration of various information sources in different areas.

---

[2]   In Iraq, for example, residents of the predominantly Shiite, Sunni, and Kurdish governorates have distinctly different concerns and attitudes on many issues.

Also captured in Figure 3.1 is the increasing dynamism as one moves from the bottom (most static) to the top layer (most dynamic). Terrain and infrastructure (the bottom two layers) generally change only very slowly (if at all), whereas the movements of people (layer three) and the changes in administrative and other boundaries (layer four) are somewhat more common. The top layer, information, is in a constant state of flux, however, with an ever-changing mix of new messages competing for attention at any given time, whereas attitudes and beliefs (layer five) change, but certainly not to the degree seen for information.

This first, geospatial lens for understanding commanders' information needs in the information domain seems to capture a number of critically important features of the information domain that were identified in our interviews and literature reviews. In fact, we presented this framework to the commanders with whom we met, and were pleased to find that almost all of them liked the taxonomy we had developed for characterizing the layers of a given information domain. Indeed, several mentioned that our layers strongly resembled some of the G2/S2 templates that were employed in Army tactical operations centers during their recent deployments. This resonance with emerging practice in the field suggests that this lens might easily be fleshed out to refine existing doctrine for military intelligence collection and analysis to better meet commanders' needs in IO and influence operations.

Network-Oriented Information. A second lens for unpacking the information domain of the operating environment characterizes features of this domain in terms of overlapping or interlocking networks. Our research suggests that conceiving of aspects of the information domain in terms of networks can offer a number of important insights. (Appendix E provides an overview of how social network analysis can contribute to influence operations.)

At the most basic level, most telecommunications, electricity, and other utilities can be conceived in terms of terrestrial or virtual networks, and many of these networks are closely associated with other phenomena of interest for IO and influence operations. We offer the following two examples as illustration.

One of the recurring complaints from Iraqis since the overthrow of Saddam and the subsequent occupation of Iraq, especially in Baghdad, is the lack of reliable electricity. Thus, Iraqis' opinions of the Iraqi government's (and coalition's) effectiveness really have turned on their assessments of the government's (and coalition's) ability to restore and protect the electricity grid or find alternatives (e.g., local generators). As a result, an area of ongoing concern for CMOs has been the repair or reconstruction of and the protection of power generation and transmission capabilities—networks easily represented and analyzed as networks of nodes and links.

A very different example can be found in the U.S. coalition's initial difficulties in grasping the importance and basic structure of Iraqi tribal networks, and the way authority and influence flowed through these networks in the post-Saddam period. When Saddam was deposed and the various governmental, patronage, and illegal networks that had supported his reign were dismantled, ordinary Iraqis fell back on a number of traditional networks, including tribal and religious networks. However, it is not clear that the key nodes in these networks were identified at an early stage in the occupation, and there seem to have been great difficulties in differentiating between the real power brokers in these networks and the various pretenders who were simply after coalition largesse. While great efforts appear to have been made to map the networks of high-value targets (HVTs) and terrorist or insurgent groups, U.S. efforts to build support for the coalition might have seen greater success if the coalition had been able to more quickly construct a picture of the key networks of power and authority in post-Saddam Iraq.

Many of the features identified by commanders with whom we spoke and in the literature we reviewed point to the conclusion that a network approach to characterizing the information domain can help to illuminate essential features of the broader political society—including key leaders, their critical relationships, and their sources of authority, power, and influence—that are of interest for IO and influence operations. Networks can be used to characterize a host of formal organizations or hierarchies, whether they are political, military, bureaucratic, or administrative; economic or business oriented; or tribal, religious, or sectarian. They also can be used to characterize informal networks, including personal and professional networks, and networks based on patronage relationships or criminal enterprises, jihadist discourse, or influence. Finally, as described above, physical networks, such as telecommunications; command, control, communications, and computers; and utilities—all of these translate naturally into link and node data.

Moreover, the ready availability of tools for portraying and analyzing networks, whether they are physical or social in nature, suggests that characterizing some key features of the information domain in terms of networks may help commanders and their staffs make sense of data that are highly relevant to IO and influence operations but would otherwise be very difficult to collect, maintain, and assess.

Group-Level Information. As discussed earlier in this report, a key feature of effective IO and influence operations is target audience analysis, a process that entails identifying which groups or audiences need to be targeted, whether that means the goal is to inform, influence, cultivate, or incapacitate them. The first step in the process is to identify the distinct stakeholder groups that may affect mission accomplishment.

To identify key stakeholder groups, a form of center-of-gravity analysis is required. One first needs to identify the most important issues being contested and the locus or forum in which the issues will be settled—e.g., executive policy councils, the legislature, the court of public opinion, or the battlefield.

One then needs to identify the key stakeholder groups or factions that will be seeking to influence the policy debate and outcome, and their overall capabilities, in terms of such factors as raw numbers of followers that can be mobilized in the streets, available economic or political resources, and men at arms or order of battle.

For each group or faction, some understanding of its identity is needed, including a characterization of its general worldview, as well as its specific aims, grievances, motivations, intentions, morale, and basic strategies. One also needs some sense of how important the issue is to the group relative to other issues on its agenda. Closely related are the distinctive attitudes, beliefs, cultural symbols (e.g., tropes to which the group will react positively, violations of cultural norms or shibboleths that are likely to result in disaffection), and historical experiences likely to constrain the ability of the United States to inform or influence the group, and the languages used. Also of interest is whether stakeholder groups are formal members of the forum where the matter will be settled or constitute outside pressure groups seeking to influence the outcome. Finally, one needs to identify each stakeholder group's or faction's key leaders and the underlying leadership and organizational structure or networks characterizing the internal hierarchies and flow of authority and power, as well as the level of cohesion or factionalization. In the case of groups with military or paramilitary wings, a sense of their strategy, doctrine, and TTP also may be of interest. A group's basic economic and other circumstances (e.g., income relative to other groups, unemployment, standing in society) also are of interest.

In a sense, the laundry list of items just described is a target folder for each stakeholder group of interest, capturing the most important characteristics of each group relevant to influence operations enterprises.

Individual-Level Information. In the best of all possible worlds, IO or influence operations would succeed on the basis of successfully informing or persuading a single individual (the head of government or head of state, for example) to behave in a certain way. In any political society, there typically are many individuals—key leaders in government, the military, religious or tribal groupings, the media, and other stakeholder groups or power brokers—who for reasons of personal influence, command of resources, leadership of large numbers of followers, etc., have a central role in influencing developments and therefore need to be directly or indirectly courted or influenced. Whether IO or influence operations are targeted against one individual or many, their success may require the collection and maintenance of a number of other types of information that, taken together, provide the information needed to plan influence operations at the individual level. These can include

- biographical materials constituting a personal history of the individual
- psychohistories and psychological profiles identifying key features of the individual's personality or personal history that may predispose him to certain sorts of behaviors or that might be exploited[3]
- analyses of decisionmaking style, including the use of information and advice, confidantes and counselors, group influences and deliberations, constraints imposed by laws or institutions, etc.
- in some cases, actor-specific models of decisionmaking that can capture the full range of cognitive, psychological, and group decisionmaking factors that may influence an individual's decisions
- preferences and viewing habits regarding TV, radio, newspaper, and other mass media
- identification of family, clan, or tribal members; members of patronage, criminal enterprise, or other networks; and known associates, as well as rivals and enemies
- connections to such institutions as businesses, banks, and criminal or other enterprises
- terrestrial, wireless, and satellite telephone numbers; email addresses, Internet service providers, and Internet protocol addresses used; and typical radio frequencies (or couriers) used for communicating with others
- the specific type of computer, cellular telephone, and other hardware used, as well as the operating systems used on these devices
- matters related to the individual's location and security, including connections to places such as residences, properties owned,

---

[3]   For an excellent analysis of actor-specific behavioral models of adversaries, see Alexander L. George, "The Need for Influence Theory and Actor-Specific Behavioral Models of Adversaries," in Barry R. Schneider and Jerrold M. Post, eds., *Know Thy Enemy: Profiles of Adversary Leaders and Their Strategic Cultures*, Maxwell Air Force Base, Ala.: United States Air Force Counterproliferation Center, November 2002, pp. 271–310.

restaurants, and other habitual locations, as well as bodyguards and movement patterns.

Resources Available to the Commander

Finally, our interviews and other lines of our analysis suggest that commanders need an understanding of the resources available to them, how they are being used, and their effects.[4] These resources include forces and other resources that are under the command of higher echelons or in adjacent AOs and may impact the operations, forces, and resources being assigned to the commander, as well as forces and resources that are assigned to subordinate commanders.

As suggested above, three primary types of information are of interest to commanders: information on the specific numbers and types of assets and capabilities available to them; information on the activities and performance of these capabilities (MOPs); information on their effects (MOEs, or outcomes). These various metrics are described in greater detail in Chapter Four and in Appendix C.

Chapter Conclusions

This chapter has summarized our thinking on how the Army might sensibly organize the various discrete types of information, as identified by commanders and the study's various other lines of inquiry, to facilitate the collection and analysis of intelligence and other data that can inform the planning, execution, and assessment of effective IO and influence operations.

We discussed all three major categories of commanders' information needs—commanders' guidance; the operating environment, including the information domain; and the resources available to the commander—but devoted most of the chapter to elaborating on a new

---

[4]   This is captured well in METT-TC, which includes a focus on mission, enemy, terrain and weather, *troops available*, and civilian considerations.

framework that the Army can use for collecting, organizing, and ana-lyzing information directly related to the information domain of the operating environment. This framework, which consists of geospa-tial-, network-, group-, and individual-level "lenses" that seem to cap-ture very well many or most of the critical features of the information domain of the operating environment, can provide the Army with both a sound approach for collecting data on these features and a platform for further systematic analysis using appropriate tools, including, for example, geographic information systems or SNA approaches or tools.

By offering some structure and precision, moreover, the frame-work can help the Army escape the somewhat vague constructions of "cultural environment" and "cultural intelligence" that seem not only to dominate discussions about commanders' information needs for influence operations, but also to have led some to conclude that cul-tural anthropology (or some other narrow field of the social sciences) offers the "silver bullet" for tackling this problem.[5] To be clear, we want to emphasize that while the framework presented here seems likely to lead to a fairly complete approach for characterizing and diagnosing the information domain of the operating environment, it also highlights the complex and interdisciplinary nature of the information, data, and analyses needed to explain the phenomenology that commanders must understand to plan, execute, and assess effective influence operations. While there may be no silver bullets, there certainly are more—and less—productive ways to approach the challenge of providing com-manders with the information they need to conduct effective influence operations.

It is critically important to differentiate any characterization of the "cultural environment" (or, in our usage, the "information domain of the operating environment") and the "cultural intelligence" of indi-

---

[5] As an intelligence analyst recently put it, "The Intelligence Community lacks a system-atic framework for fully understanding what 'cultural intelligence' means." Quoted by Julia Riva in her review of P. Christopher Earley and Soong Ang's *Cultural Intelligence: Individual Interactions Across Cultures* (Palo Alto, Calif.: Stanford University Press, 2003), in *Studies in Intelligence*, Vol. 49, No. 2, 2005.

viduals.[6] As stated earlier, our research did not address the question of what *de minimis* level of training and education most deploying soldiers should receive to improve their "cultural intelligence."[7] Nor did we address which specific branches or specialties (e.g., foreign area officers, Special Forces, PSYOPers, public affairs specialists, political advisors, FA-30 IO officers) should be commanders' principal advisors on cultural and related matters, much less how *they* should be educated, trained, or equipped. It is worth noting, however, that many commanders' information needs in this arena actually might be best met by having highly capable advisors who are steeped in the local color, language, and traditions. This matter, however, is necessarily the focus of a very different study from the one we did.

Before concluding this chapter, we turn briefly to three important issues pertaining to the collection of data on the information domain of the operating environment.

The first issue is the question of what types of information should be collected and when. We think that to avoid creating databases filled with highly perishable information whose utility may erode before its use, the emphasis during peacetime should be on collecting or assembling information on the relatively static aspects of national and subnational entities that are implicated in operational or contingency planning. This would include information at the national level (e.g., demographic or economic in nature) or related to geospatial characteristics or infrastructure, as well as basic information on key groups and leaders, existing media and public opinion data, etc. Responsibility for collecting or assembling this information should fall to Army G-2, NGIC, 1st IOC, 4th PSYOP Group, regional combatant commands, Defense Intelligence Agency, and other members of the intelligence

---

[6]   Earley and Ang define cultural intelligence as "a person's capability to adapt to new cultural contexts" (Riva, 2005).

[7]   The Army currently has programs for providing soldiers deploying to Iraq and Afghanistan with a basic level of understanding of cultural matters.

community; and as appropriate through transference to selected academic/other studies and/or database development efforts.[8]

During the transition to operations, the focus increasingly should turn to more highly perishable information, as well as information that is impossible to collect during peacetime. Responsibilities for data collection during operations necessarily will fall to some combination of military capabilities in the field—with the G-2/S-2 taking the lead, but "every soldier a sensor" reporting graffiti, smiles, hand waves, the results of face-to-face encounters while on patrol, etc.; and with frequent supplementation from field support teams and contractors that can conduct focus groups, surveys, media content analyses and provide other support—and to reachback capabilities to the institutions taking the lead during peacetime.

Moreover, important decisions will have to be made about which systems will become the standard ones for capturing, tracking, and supporting analysis of these various data, who will have the responsibility for maintaining and updating these data, and a host of other issues that are well beyond the scope of this report. Nevertheless, based as they are on commanders' own views about the information they need to conduct effective influence operations, the taxonomies offered here should provide the Army with a reasonable starting point for organizing, training, and equipping soldiers to provide their commanders with the information they need.

---

[8] Indeed, the 4th PSYOP Group produced an IPB summary before OIF that in many respects resembles our framework.

CHAPTER FOUR

Remaining Challenges

As we stated at the outset, we lack a firm basis for establishing the prevalence in the field of many of the problems that were identified by the commanders. Nevertheless, the fact that certain issues were mentioned by multiple commanders suggests that they clearly deserve closer examination by the U.S. Army. In this concluding section, we discuss four emerging challenges in meeting commanders' information needs and conducting effective influence operations. Although these challenges are based largely on anecdotal evidence, we think they are important enough to deserve attention, further analysis, and potential remedial action.

Vertical Coordination and Echelonment

A recurring theme from our research was the requirement for integrated planning, execution, assessment, and information flows between echelons to ensure complementarity and synergy in influence operations.[1] There is some evidence from our commanders' interviews and our review of the published record, however, that brigades in the field have encountered challenges in this regard.

---

[1] This point is stressed in, among other sources, Gary J. Schreckengost and Gary A. Smith, "IO in SOSO [Stability Operations and Support Operations] at the Tactical Level: Converting Brigade IO Objectives into Battalion IO Tasks," *Field Artillery*, July–August 2004, pp. 11–15.

Some brigade- and battalion-level personnel with whom we spoke affirmed the increasingly important role of brigade-level influence operations in the field. They also noted some emerging difficulties in ensuring complementarity and synergy between brigade- and corps-level influence operations, as well as between brigade and battalion operations.

Our interlocutors said that corps-level IO and influence operations messages frequently failed to make necessary distinctions between target audiences in their AO, or to focus on specific desired behaviors. As a result, these messages often failed to resonate with specific groups in the AO because they were too generic and diffuse, or they failed to achieve a desired effect because none was clearly specified.

Another concern our interlocutors raised about brigade-corps operations had to do with significant delays encountered when seeking approval for brigade-level messages. A recurring problem in the field appears to be that by the time the corps approves a brigade's message, the situation on the ground in the brigade's AO has changed, and the original recommended message has been overtaken by events. There is at least some evidence that as a result, brigades increasingly are conducting influence operations without benefiting from the synchronization of the IO and influence operations capabilities that reside at the corps level.

Perhaps even more important is that the commanders and former officers on battle staffs with whom we spoke suggested that the biggest challenges may lie in the battalion-brigade relationship. According to these officers, this is where disconnects between themes and messages and long approval times appear to be especially significant.

Needless to say, this issue, if unresolved, could be leading to balkanization and incompatibilities in the establishment of metrics for influence operations, where lower-echelon objectives and metrics for such operations are not adequately nested, making it impossible to produce a coherent framework for assessment from the battalion up to the corps and theater levels. A recurring challenge, therefore, will be to balance the desiderata of broad top-down planning guidance with the need for flexibility at lower echelons so as to ensure that influence efforts are tailored to local conditions.

Horizontal Coordination Across Areas of Operation

Brigade-level commanders also noted the requirement for horizontal coordination and information flow between brigade operations in adjacent AOs.[2] Our impression from our interviews and other research, however, is that the importance commanders place on coordinating their influence operations activities with commanders in adjacent AOs, and the mechanisms they use for assuring this coordination, are somewhat ad hoc in nature.

Our structured conversations with commanders and former members of battle staffs suggest that adjacent brigades in Iraq generally were coordinating well enough on messages to avoid "IO fratricide," but that synchronization of messages across adjacent AOs has become a significant challenge. According to our interlocutors, difficulties in synchronization across AOs led to messages being emphasized at different times in different sectors. This may be causing confusion among ordinary Iraqis who move across brigade boundaries or talk to relatives in other AOs and find that different messages are being emphasized; it also raises questions about what the United States' principal message might be at any given time.

Without a strong division- or corps-level process to ensure that influence operations are synchronized across echelons and subordinate AOs alike, the risk that efforts to push influence operations down to the brigade level will result in balkanization and lack of synchronization is great. How best to manage the tension between the desiderata of a top-down strategic and operational planning process and timely, responsive, and effective tactical implementation seems likely to remain an ongoing challenge for commanders.

Ensuring Continuity in Transitions

It is critically important that influence operations minimize abrupt changes that may confuse or increase the uncertainty or fears of the

---

[2]   See, for example, Baker, 2006, pp. 13–32.

target audience. The operational implication is that significant efforts should be made to ensure not only continuity in the application of influence operations across brigade rotations, but also the availability of operation-relevant information across rotations. As stated in a recent article about intelligence in COIN operations that applies with equal force to influence operations:

> Battle handover between units must not disrupt continuity. Processes must be in place to ensure analysts moving into a theater are able to understand the intelligence picture, the intelligence plan, and applicable intelligence databases. Without continuity, the intelligence picture will begin anew with every troop rotation, and there will be no consistent long-term analysis of the insurgency.[3]

Our interviews with commanders suggest that current efforts to ensure smooth transitions between units, and thereby enhance a sense of continuity in influence operations, may be inadequate and may need to consist of much more than a "right-seat ride."

For example, we heard of cases in which the influence operations of a unit being relieved were immediately jettisoned by the new unit's commander greater emphasis on kinetic activities, such as cordon-and-search operations. Equally important, brigades currently lack a common set of databases and tools that might help to ensure continuity by providing newly arriving units with a sense of the earlier unit's history in the AO, and what mix of IO, influence operations, and kinetic activities the earlier unit found to be most effective or ineffective in that AO. Put another way, it is the opinion of some commanders that rather than building on lessons learned based on multiple unit rotations in an AO, each unit has tended to rotate in and make changes without a full appraisal of what elements of IO and influence operations might already be working.

The implication for the Army is that additional efforts and mechanisms are needed to provide units that are rotating in with an endow-

---

[3]   Kyle Teamey and Jonathan Sweet, "Organizing Intelligence for Counterinsurgency," *Military Review*, September–October 2006, pp. 24–29.

ment of relevant experiential information—chronologies, network analyses, contact files, databases, and other types of information—that can assist a new commander in understanding the history and authority structures of the AO. For example, mechanisms can be developed to assist units in "shadowing" the activities and battle rhythm of the unit they are relieving in the months leading up to a deployment, or to treat the deploying unit as a reachback capability that can assist in assessing influence operations.

Moreover, it is not clear that the incentives commanders face support the sort of continuity needed for effective influence operations. The promotion system creates great incentives for battalion and brigade commanders to innovate and implement a brand new influence strategy in their AO when they arrive; commanders need to show that they "made their own mark" on their AO during their rotation in order to do well in their officer's evaluation reports. This can make it more difficult for transitions to be smooth, no matter how much money is invested in digital right-seat-ride technology. Rather than rewarding commanders for changes to their predecessor's influence operations, it might be desirable to reward them for improvements on relevant metrics that span multiple deployments.

Overcoming Doctrinal Stovepiping of Information Operations

The final challenge is what we see as a necessary doctrinal shift, moving from a joint and Army conception of IO and influence operations as a set of discrete stovepipes to a conception focused more on the contributions of these operations to achieving the objectives of combined arms, joint, and combined operations.

Our interviews and other lines of analysis suggest that IO success in the field increasingly depends on commanders' ability to think beyond the procrustean bed of current IO doctrine, which tends to focus on the employment of IO in major combat operations and treats IO and its related and supporting capabilities as discrete disciplines rather than capabilities whose employment needs to be planned, synchronized, and executed in concert with the other combined arms to produce desired effects and outcomes.

Given that the latest release of JP 3-13, *Information Operations*, was in February 2006, it seems highly unlikely that JP 3-13 will be rewritten any time soon. However, the Army's FM 3-13, *Information Operations: Doctrine, Tactics, Techniques and Procedures*, is dated November 2003 and was written before this release of JP 3-13. Moreover, CAC's recent thinking about IO is not reflected in the current Army doctrine. Thus, it can be argued that a rewrite of FM 3-13 is in order.

We think that Army IO doctrine should be revised to emphasize its employment in a wider range of operations, from COIN and stability operations to major combat operations. In particular, FM 3-13 should go beyond its focus on offensive and defensive IO to consider IO in other contexts, such as COIN and stability operations. We also think that the Army should consider including in selected mission-specific doctrinal publications (e.g., FM 3-24, *Counterinsurgency*) a chapter or appendix detailing specific considerations pertaining to the effective employment of influence operations in that mission context. Education and training programs also will need to be revised to train future commanders on the principles of employing influence operations across a wide range of mission types. In the end, we believe doctrine, education, and training should capture best practices from the field in integrating influence operations into combined arms operations.

Our assessment of commanders' information needs for influence operations has led us to conclude that influence operations should be treated like all other operations in the battle space: They should be well informed by intelligence collection and analysis of the population and its information domain; they should be coherently planned in a top-down fashion enabling them to be integrated with joint and combined actions and with non-kinetic LOOs; they should be executed at all levels in a way that provides for vertical, horizontal, and temporal consistency, as well as an orderly transfer of authority; and they should provide mechanisms for monitoring inputs, outputs, and outcomes so that operations can be adapted to changing circumstances. As described in this document, to achieve these desiderata may require additional changes to existing doctrine and organizations and to training and educational programs.

Identified Information Requirements for Influence Operations

This appendix provides a detailed list of the information requirements for influence operations that were identified over the course of our study and only summarized in the main text.

Indicators from the National Training Center

Indicators for Red Activities

Who/where are the anti-Iraqi force (AIF) leaders?

- What groups and personalities are working to undermine coalition stability objectives?
- Which villages or groups are providing safe haven and support to insurgents or terrorists?
- What is the AIF hierarchy?
- Who are the financiers?
- Who are the organizers and C2?
- Who are the IED/vehicle-borne IED (VBIED) manufacturers?
- Are the different AIFs cooperating or competing?
- What are locations of C2 and key AIF leaders; location of crossing sites, location of weapons caches, supplies? What are the enemy's attack patterns in time, space, effect?
- Which Sunni or Shi'ite institutions or leaders are supporting insurgent activities in the BCT AO?

What are incidents/indicators of AIF psychological warfare activities, capabilities, intentions?

- Where are the locations of C2, logistic nodes, and communication nodes?
- Where in the AO are the Internet satellite dishes?
- Where are the enemy propaganda banners and flyers being produced?
- Where are the high-volume printing/copying outlets?
- Describe the radio station(s)—frequency, footprint, grid of antenna and station, format, bias, owner/key personnel.
- Where are large concentrations of blank CDs and DVDs (enemy IO video reproduction)?
- Describe the newspaper(s)—name, bias, frequency, distribution locations, circulation numbers.
- Is there an increase of propaganda/protests?
- Where are enemy propaganda flyers and banners appearing?
- Will AIF incite anti-American/Iraqi demonstration in order to disrupt 3rd interim brigade combat team movement?
- Is anyone planning or executing a civil disturbance?
  Miscellaneous Red activities:
- What is the reaction of AIF to the 3/2 Stryker brigade combat team (SBCT) entry into AO ARROWHEAD?
- Are AIFs increasing attacks against friendly forces?
- What foreign influences are acting within our AO?
- How does the Iranian Intelligence Service influence MM in our AO?
- How will AIF try to intimidate the command, control, communications, and intelligence (C3I) system?
- How is AIF recruiting in the AO?

Indicators for Sewer, Water, Electric, and Telecommunications infrastructure

- What is the status of critical facilities and services needed to accomplish the 3-2 SBCT SSTR objectives?

- What are the reductions in services to populace—security, medical, informational, and transportation?
- Are any essential services targeted by AIF (water tankers, water sources, food supplies, banks, schools, generators, etc.)?

Indicators for Local Elites

Who requires protection?

- Pro-government and Iraqi Police Service leaders?
- Who are the publicly identified informants?
- Who are effective providers of resources to the communities?
- What necessary community resources (heating, electricity, water, etc.) are threatened by the insurgents?
- Where are the police stations in our AO?
- Will (name of particular leader) become pro–multi-national forces (MNF)?
- Are there outside police/government officials connected to police officials in the affected areas that we can influence?

Indicators for Attitudes of General Population

- Where are the media outlets in the AO?

Indicators from Selected Military Journal Articles

Indicators from GEN David H. Petraeus, "Learning Counterinsurgency: Observations from Soldiering in Iraq," *Military Review*, January–February 2006, pp. 2–12:

- Understand the half-life factor, length of time before a counterintelligence force becomes seen as an army of occupation.
- Number of small infrastructure repair projects being undertaken in direct response to local demand.
- Is the number of local stakeholders in operational success increasing or decreasing?

- To what extent are kinetic operations creating more recruits for the insurgent cause?
- How extensively developed are the local HUMINT networks?
- Estimate level of respect for U.S. forces' cultural awareness in dealing with local leaders and civilians.
- Number of damaging tactical incidents caused by mistakes by strategic corporals.

Indicators from LTG Thomas F. Metz et al., "Massing Effects in the Information Domain: A Case Study in Aggressive Information Operations," *Military Review*, May–June 2006, pp. 2–12:

- Where is the IO threshold at which bad publicity cripples kinetic operations?
- How can that threshold be raised?
- Who will be the key local "influencers" during a given operation?
- Which are the key centers of Red misinformation dissemination?
- How can Blue best collect and disseminate photographic documentation of insurgent atrocities?

Indicators from GEN Peter W. Chiarelli and MAJ Patrick R. Michaelis, "Winning the Peace: The Requirement for Full-Spectrum Operations," *Military Review*, July–August 2005, pp. 4–17:

- What are the ethnic, religious, and cultural factors that dominate the AO?
- Who are the key local Iraqi facilitators and stakeholders?
- What is the status of the local infrastructure?
- What do the key indicators of economic progress (i.e., prices, wages, unemployment rate, waiting times at gas pumps) tell us?
- Plot correlations between insurgent attacks and areas of weak infrastructure.

Indicators from COL Ralph O. Baker, "The Decisive Weapon: A Brigade Combat Team Commander's Perspective on Information Operations," *Military Review*, May–June 2006, pp. 13–32:

- Who are the key local power brokers?
- Which are the most popular newspapers and TV stations?
- Number of favorable versus unfavorable reports on Arab satellite TV networks.
- Trends in the number of intelligence tips received from local civilians.
- Changes in the tenor of Friday mosque sermons.
- Status of reconstruction projects.
- Amount of damage to vital local infrastructure caused by insurgent attacks.

Indicators from Structured Conversations with Commanders

- Understand the cultural dynamics of the country within which you are operating.
- What are the attitudes of the major demographic groups in a country (e.g., rural Sunnis, urban Shiites, urban Kurds)?
- What are the ideological/theological weaknesses in specific jihadist IO messages?
- Which moderate Islamic clerics could effectively counter jihadist IO?
- How do IED events harm the local civilian population?
- What is the delta between the expectations of the local civilian population and the economic/infrastructure realities on the ground?
- How is the tenor of Friday mosque sermons changing over time?
- What are the trends in the number of good intelligence tips received from the local civilians?
- Determine the location of the IO threshold (i.e., the point at which bad publicity halts a kinetic operation).
- Tone of reporting on U.S. activities in local newspaper and on local TV stations.
- Number of civilians antagonized by poorly targeted raids.

- Timeliness of Blue IO messages—were they relevant to the locals?
- Who are the key power brokers in the local population (e.g., sheikhs, imams, schoolteachers)?
- Which TV networks are the most watched?
- Number of ongoing reconstruction projects with strong symbolic value.
- Number of local figures willing to propagate U.S. themes and messages.
- Number of religious figures who pre-approve targeted mosque raids.
- Level of local perception that United States is trying to limit civilian casualties.
- Level of success of local elections.
- Amount of alignment between word and deed—are promises being kept by Blue?
- Attitudes of local NGOs toward U.S. activities.
- Local views of U.S. treatment of detainees.
- What are Red's main IO themes and messages?
- What is the "on the street" behavior of the locals toward U.S. patrols (e.g., friendly waves or obscene gestures)?
- How is the amount of anti-American graffiti changing over time?
- Who are the most trusted sources of information in the local area?
- Polling data, survey on local attitudes conducted by local survey organizations.

Task List Analysis

Part of our effort to identify commanders' information needs for IO and influence operations entailed analyzing a task list to identify the tasks related to IO and influence operations that are assigned to the Army and to assess the relationships between those and other tasks that may be needed to influence targeted individuals and groups. Chapter Two summarizes this portion of our effort; this appendix explains how the task list analysis was performed and what it revealed.

Taxonomy of Information Operations Tasks and Effects

A number of discrete effects may be sought in IO and influence operations, and they are generally divided into groups of offensive effects, defensive effects, and effects that seek to inform and influence.[1]

Offensive IO Tasks and Effects

- *Destroy:* To damage a combat system so it cannot perform any function or be restored.

---

[1] The main sources for this taxonomy are Headquarters, Department of the Army, 2003c, pp. 1-16 to 1-17; 1st IOC (Land), Field Support Division, "Terminology for IO Effects," in *Tactics, Techniques and Procedures for Operational and Tactical Information Operations Planning*, March 2004, p. 23; and Schreckengost and Smith, 2004, pp. 11–15. Entries noted as not further defined are mentioned in Schreckengost and Smith, 2004, as "traditional IO tasks."

- *Disrupt:* To break or interrupt the flow of information between selected C2 nodes.
- *Degrade:* To use non-lethal or temporary means to reduce the effectiveness or efficiency of adversary C2 systems and information collection efforts.
- *Deny:* To withhold information about Army force capabilities and intentions that adversaries need for effective and timely decisionmaking.
- *Deceive:* To cause a person to believe what is not true.
- *Exploit:* To gain access to adversary C2 systems to collect information or to plant false or misleading information.
- *Isolate:* To prevent an enemy leader or unit or a populace group from communicating with others.

Defensive IO Tasks and Effects

- *Detect:* To discover or discern the existence, presence, or fact of an intrusion into information systems.
- *Protect:* To guard against espionage or capture of sensitive equipment and information.
- *Restore:* To bring an information system back to their original state.
- *Respond:* To react quickly to an adversary's information operations attack or intrusion.
- *Mitigate:* To reduce the effects of an adversary's operations in the information environment.
- *Preserve:* To maintain the effectiveness or efficiency of friendly force information systems, assets, or functions.

Inform and Influence Tasks and Effects

- *Influence:* To cause adversaries or others to behave in a manner favorable to Army forces.
- *Promote:* To increase acceptance of an idea, concept, event, activity, or operations.

- *Inform:* To provide information to or educate a specific target audience.
- *Encourage:* Not further defined.
- *Divert:* Not further defined.
- *Warn:* Not further defined.

Task List Analysis

All tasks for which the Army is formally responsible are documented in two principal sources: *Universal Joint Task List* and *The Army Universal Task List.*[2] Therefore, the analysis began with a search of these two documents for all references to functions having to do with IO—i.e., functions that in some way perform, support, or depend on IO. We compiled two lists of associated tasks, one drawn from the UJTL and one drawn from the AUTL. Then we searched the source documents for additional, non-IO tasks whose descriptions had to do with influencing any individuals or groups—friendly, adversarial, or other—and added those tasks to the lists.

Next, using joint and Army IO doctrine publications as guides, we identified the aspects of IO that each task supports and coded tasks in terms of whether joint and Army doctrine considers those functions to be core or supporting elements, or merely "related activities."[3] As Table B.1 illustrates, JP 3-13 and FM 3-13 are both reasonably consistent in the ways they describe IO; but as this analysis demonstrates, neither quite captures all the elements that support IO or activities related to influence operations.

---

[2]   CJCSM 3500.04D (DoD, 2005) and FM 7-15 (Headquarters, Department of the Army, 2003b).

[3]   Joint and Army doctrines are provided, respectively, in JP 3-13 (DoD, 2006b) and FM 3-13 (Headquarters, Department of the Army, 2003c). Terms are not always consistent in the two doctrines: FM 3-13 refers to core and supporting elements of IO and related activities, whereas JP 3-13 refers to the same functions as core, supporting, and related capabilities. For clarity, we use Army terms except when referring specifically to joint doctrine.

Table B.1
Core, Supporting, and Related Functions of IO and Influence Operations

Joint Doctrine	Army Doctrine
Core Functions	
Electronic warfare	Electronic warfare
Computer network operations	Computer network operations
	Computer network attack
	Computer network defense
	Computer network exploitation
PSYOP	PSYOP
Operations security	Operations security
Military deception	Military deception
Supporting Functions	
Information assurance	Information assurance
Physical attack	Physical destruction
Physical security	Physical security
Counterintelligence	Counterintelligence
Combat camera	
	Counterdeception
	Counterpropaganda
Related Functions	
Public affairs	Public affairs
Civil-military operations	Civil-military operations
Defense support to public diplomacy	

According to JP 3-13 and FM 3-13, the core capabilities of IO are electronic warfare; computer network operations (CNO), including computer network attack (CNA), computer network defense (CND), and computer network exploitation (CNE); PSYOP; operations security; and military deception.[4] Joint doctrine lists the supporting capabilities as information assurance, physical attack, physical security, counter-

_____

[4]   Oddly, FM 3-13 (Headquarters, Department of the Army, 2003c, p. 2-9) repeatedly lists CNO, CNA, CND, and CNE as coequal core elements of IO and yet states that "computer network operations *comprise* computer network attack, computer network defense, and

intelligence, and combat camera. Current Army doctrine names these functions, except for combat camera, as supporting elements and adds counterdeception and counterpropaganda. According to joint doctrine, public affairs, CMO, and defense support to public diplomacy are capabilities related to IO. Current Army doctrine lists only public affairs and CMO as related activities. Although neither joint nor Army IO doctrine lists either ISR as a supporting element or civil affairs as an activity related to IO, we included them in this analysis because of ISR's importance for providing information to commanders and civil affairs' potential impact on influence operations.

Tasks were also coded according to which of six basic IO-related or influence operations–related functions they perform:

- *Policy:* Tasks relaying political guidance or providing senior military guidance to joint forces on IO or influence operations.
- *C2:* Tasks that involve planning, managing, synchronizing, or coordinating IO-related or influence operations–related functions.
- *Inform:* Tasks that provide information needed to perform IO or influence operations.
- *Offense:* Tasks that primarily employ offensive IO-related or influence operations–related functions. Offensive functions include destroy, degrade, disrupt, and deny; deceive; exploit; and influence (when referring to efforts to influence enemy forces rather than other target audiences).
- *Defense:* Tasks that primarily employ defensive IO-related or influence operations–related functions. Defensive functions include protect, detect, restore, and respond.
- *Influence:* Tasks that emphasize influencing target audiences other than enemy forces.

Once all tasks related to IO and influence operations were identified and coded, we tallied tasks in each category and assessed the

---

related computer network exploitation enabling operations" (emphasis added), implying a hierarchical relationship with CNA, CND, and CNE as components of CNO.

results to determine what the Army's IO and influence operations tasking emphasizes. The results were informative. Figure B.1 illustrates how IO-related and influence operations–related tasks in the UJTL and AUTL are proportioned between core elements, supporting elements, and related activities.

Perhaps the most striking observation from Figure B.1 is the essential complementarity of supporting and related tasks, which comprise anywhere from about one-half to two-thirds of the tasks (core IO tasks comprise about one-third to about one-half of the tasks). We also were struck by the completeness of the task lists—they seem to capture all key activities that we would associate with IO and influence operations. Finally, and somewhat impressionistically, the task lists also seem to specify or imply commanders' information needs in a relatively complete way.

In the UJTL, functions related to core elements of IO were the majority of IO-related and influence operations–related tasks identified. Those related to supporting elements were the second most frequently seen, with tasks associated with related activities appearing less often. In the AUTL, the majority of relevant tasks were found in areas

Figure B.1
Basic Distribution of Tasks Related to IO and Influence Operations

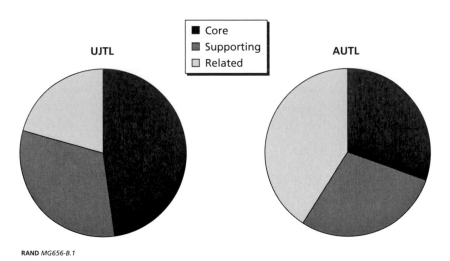

that IO doctrine considers only related activities, with the remainder of tasks about evenly distributed between core and supporting elements of IO. A closer examination depicts this contrast more clearly. It also reveals a peculiar disconnect between the majority of core IO tasks and those directly associated with influencing individuals and groups typically targeted in stability and COIN operations. Figure B.2 illustrates how the tasks related to IO and influence operations are distributed according to the six basic functions of IO identified above.

As would be expected, the UJTL contains a greater proportion of tasks dedicated to force management than does the AUTL, with about one-quarter of them pertaining to policy and C2 functions. The UJTL also has a somewhat higher proportion of tasks involving offensive and defensive IO than does the AUTL, as well as a greater proportion of tasks devoted to providing commanders with the information needed to perform IO and influence operations (tasks coded "Inform").

What is most notable, however, is that the AUTL has a much higher proportion of tasks associated with influencing individuals and

Figure B.2
Distribution of Functions Related to IO and Influence Operations

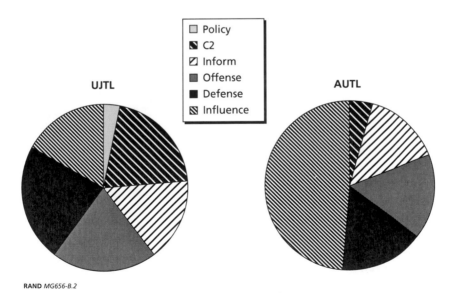

groups other than enemy forces. This is significant, because these are the actors whose cooperation is most critical to the success of stability and COIN operations. But ironically, the tasks most relevant to influencing these actors were least likely to be associated with the core and supporting elements of IO. Rather, they were more directly addressed in functions that joint doctrine and Army doctrine consider only related activities; and an important one, civil affairs, is not even addressed in IO doctrine. Figure B.3 illustrates this peculiar inverse relationship.

As the detailed distribution of AUTL tasks reveals, the functions that joint and Army doctrine consider the core and supporting elements of IO focus almost exclusively on efforts to gain "information superiority" over enemy forces. Functions such as electronic warfare, CNO,

Figure B.3
Detailed Distribution Tasks Related to IO and Influence Operations Tasks in the AUTL

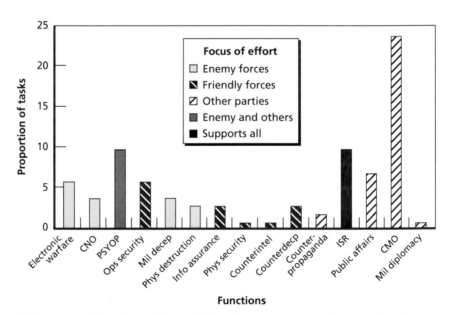

NOTE: For legibility, CNA, CND, and CNE tasks have been combined and listed as CNO. If broken out, columns aggregating CNA and CNE tasks would be light gray, and the CND column would be black with white diagonal stripes.
RAND *MG656-B.3*

military deception, and physical destruction are aimed at degrading the enemy's decisionmaking processes, while operations security, information assurance, physical security, counterintelligence, and counterdeception focus on protecting the quality and timeliness of friendly information and denying the enemy access to it.

In fact, PSYOP and counterpropaganda are the only core or supporting elements of IO that include tasks aimed at influencing actors other than enemy forces. Ironically, the overwhelming number of tasks focused on influencing individuals and groups most critical to successful stability and COIN operations reside in functions that joint and Army doctrine consider to be activities related solely to IO. Tasks supporting public affairs, CMO, and defense support to public diplomacy (labeled "Mil diplomacy" in Figure B.3) are crucial to successful influence operations. The heavy emphasis on tasks supporting CMO in the current AUTL no doubt reflects the Army's recognition of the importance of such tasks to the success of efforts in Afghanistan and Iraq. The fact that JP 3-13 and FM 3-13 categorize such functions as "related capabilities" or "related activities" suggests, however, that IO doctrine remains too heavily focused on major combat operations at the expense of providing insufficient guidance for conducting effective IO in the kinds of operations in which the Army is most heavily engaged today.

A Metrics-Based Planning and Assessment Approach for Influence Operations

One of the study team's tasks was to describe how a metrics-based planning and assessment approach for IO and influence operations that was developed for an earlier study might be implemented in combined and joint operations. This appendix provides that description.

Recent joint doctrine designates IO as a "core competency" in the military and highlights for the first time efforts to develop MOEs for IO tasks in relation to a commander's operational objective.[1] However, while doctrine indicates the need to measure outcomes, it is less clear about which kinds of assessments are needed to determine the effectiveness of combat operations and the specific contributions of IO and influence operations to the achievement of tactical and operational-level objectives.

A perennial assessment problem, and one that was beyond the scope of our effort, is the greater difficulty of establishing causation (i.e., that a specific activity caused or contributed to a specific outcome) as opposed to correlation (i.e., that an outcome was statistically associated with an activity). Wars are not controlled experiments, so efforts

---

[1]   Christopher J. Lamb, "Information Operations as a Core Competency," *Joint Forces Quarterly*, Issue 36, December 2004, p. 88. JP 3-13 (DoD, 2006a) was the first update since October 1998. The 2006 publication was the culmination of recommendations and guidance issued in several of the preceding years, including in the *2001 Quadrennial Defense Review Report*, the October 2003 DoD *Information Operations Roadmap*, and the 2004–2009 Defense Planning Guidance. See DoD, *2001 Quadrennial Defense Review Report*, September 30, 2001b; and DoD, *Information Operations Roadmap*, October 30, 2003.

to prove causation generally are fruitless. Nevertheless, under limited circumstances, e.g., where one is able to rule out all other plausible causes, one can sometimes argue that a specific activity was a contributor to a specific outcome.

To address this issue, we describe a basic approach for measuring influence operations' effectiveness and suggest how this approach could be managed within existing command functions. Figure C.1 provides an overview of the planning and assessment process as conceptualized by the study team. While planning directives are generated at higher echelons and then used to guide the actions of lower echelons, assessments are made at lower echelons and then fed up to higher echelons. A metrics-based approach to IO planning and assessment can help to support the flows of information in both directions.

Assessment of IO effectiveness can be difficult for several reasons. First, it often is not clear to units what they should be measuring. For example, it is not uncommon for one unit to judge operations on the basis of the number of leaflets dropped (an input, or MOP),

Figure C.1
Flows of Information for Planning and Assessment System

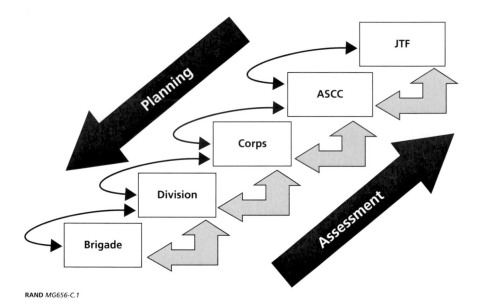

while another focuses on whether a targeted audience picked up and read the leaflet (an output measure), and still another focuses on whether the behavior of the target audience changes in accordance with the instructions on the leaflet (an MOE).

Another difficulty stems from units seeking to measure IO and influence operations only after the fact, during after-action reviews, and thus basing their assessments on whatever data might still be available. In such circumstances, assessors may need to pore over reams of data looking for something "significant"—even when the available data do not speak particularly well to IO's effectiveness—or the units may not have clear guidance on the sorts of criteria they should use. Because units do not adhere to a common standard or approach for measuring, and because differences can arise from one echelon to another or one AO to another, the results of assessments cannot easily be "aggregated up" to provide commanders with a composite common operational picture of IO and influence operations performance.

In addition to the current assessment process making it difficult for commanders to understand the contribution of IO, the current planning process does not provide units with the kind of specific direction needed to track performance in a way that stays focused on the "big picture" of the commander's intent, visualization, and operational objectives. As a result, units might track whether or not an activity was performed (e.g., leaflets distributed, a military feint attempted) but will have little guidance on how to assess the links between this activity and the commander's operational and tactical objectives or on how to assess and report whether the task achieved an intended effect. Since IO capabilities affect objectives at all levels of war and across the range of military operations, these capabilities must be consistent with broader national security policy and strategic objectives.[2]

The approach outlined here aims to address these issues.

---

[2]   JP 3-13 (DoD, 2006a), p. I-8.

Defining Metrics for Information and Influence Operations

In developing operationally useful metrics—for IO and influence operations or any other purpose—a number of considerations apply (see Figure C.2). First, one needs to identify desired end states in operational terms. One then needs to identify, for each desired end state, observable quantitative or qualitative indicators that will serve as useful proxies for measuring progress toward the desired end state, and some criteria of success (i.e., a target value for each indicator that establishes a clear threshold or yardstick for judging whether the end state has been achieved). Finally, one needs to identify the source of the data for the indicators: intelligence channels, operational units, liaisons, etc.

A Metrics-Based Planning and Assessment Methodology

Overview

To understand how the planning and assessment process can be used to track measures of IO effectiveness and underwrite the top-down planning and bottom-up assessment process suggested in Figure C.1, we need to embed such considerations in a framework that links overall campaign objectives and objectives of subordinate AOs, tasks and their effectiveness, and capabilities and their performance during the planning phase (Figure C.3).

As the figure shows, campaign or operational objectives are established by the combined or joint forces commander, the land component commander, or the corps-level organization responsible for overall conduct of the campaign. The operational headquarters translates each operational objective into a measure of operational outcome (MOO) by establishing relevant criteria of success, as well as outcome indicators and data sources that are needed to populate these indicators.

Campaign-level objectives flow down to units in subordinate AOs and are mirrored at the subordinate AOs, where planners must address the question, What tactical objective must be achieved within the subordinate AO to support the overall campaign objective?

Figure C.2
Elements of Metrics

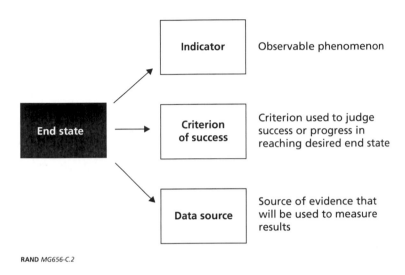

Figure C.3
Overall Flow of Metrics-Based Planning and Assessment Methodology

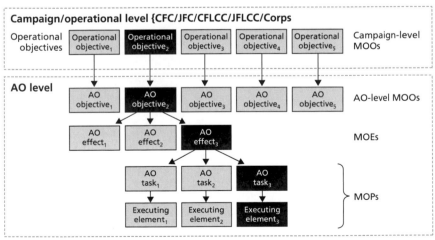

When the planning staff of a subordinate AO receives an operational objective, the staff is responsible for translating that objective into a tactical objective for the local AO. We call this local objective the "AO objective," meaning the desired end state at the tactical level for a given AO that is needed to achieve the larger operational objective. Again, relevant criteria of success, outcome indicators, and data sources are identified.

Planners must next address the question, What AO effects must be achieved to reach the local AO objective? Because one or more AO effects might be required to achieve an AO objective, planners then identify the local effects that cumulatively are judged likely to lead to the achievement of the AO objective, as well as the tasks needed to produce these effects. We call these "AO effects"—i.e., the effects needed to achieve the AO objective.

Planners next address the question, What tasks must be successfully accomplished to achieve the desired AO effect? AO effects are produced by "AO tasks"—i.e., the tasks that need to be successfully conducted within the AO to yield the desired AO effect. As suggested by Figure C.3, one or more AO tasks might be required to achieve an AO effect. Again, relevant criteria of success, outcome indicators, and data sources are identified for the AO effects.

Finally, planners address the question, What capability performance levels are needed to accomplish an AO Task? Planners establish MOPs—including indicators, criteria of success, and data sources—that specify the desired level of task accomplishment by IO elements or related activities to successfully perform the task and achieve the associated AO effect.

As just described, our framework provides a systematic way of connecting operational objectives down to individual elements and related activities in the planning process and, as the necessary data are collected, pushing these data back up the chain in a form that can be assessed in an integrated fashion.

We next provide definitions of the methodology's building blocks and then turn to an illustrative application of the methodology.

Some Definitions

The study team, building on prior work for the U.S. Army, recognizes that at the highest level, the key assessment in any operation is whether overall operational objectives have been achieved.[3] The achievement of these objectives depends in turn on whether planned tasks were properly executed and the effects can be attributed to those tasks. Before proceeding with a detailed application of our approach, we define some essential terms:[4]

- *End state:* At the operational and tactical levels, the conditions that, when achieved, accomplish the mission. At the operational level, the conditions that attain the aims set for the campaign or major operation. (FM 3-0)[5]
- *Criteria of success:* Information requirements developed during the operations process that measure the degree of success in accomplishing the unit's mission. They are normally expressed as either an explicit evaluation of the present situation or forecast of the degree of mission accomplishment (FM 6-0). In our usage, criteria of success connote the target or threshold levels of a metric that must be achieved or exceeded for a commander to declare that an objective has been achieved or that the desired effect of a task has been realized.
- *Evaluation:* Comparison of the relevant information or metric on the situation or operation against criteria to judge success or prog-

---

[3] Past RAND work on metrics includes Jefferson P. Marquis, Richard E. Darilek, Jasen J. Castillo, Cathryn Quantic Thurston, Anny Wong, Cynthia Huger, Andrea Mejia, Jennifer D. P. Moroney, Brian Nichiporuk, and Brett Steele, *Assessing the Value of Army International Activities*, MG-329-A, Santa Monica, Calif.: RAND Corporation, 2006; and Walter L. Perry, Robert W. Button, Jerome Bracken, Thomas Sullivan, and Jonathan Mitchell, *Measures of Effectiveness for the Information-Age Navy: The Effects of Network-Centric Operations on Combat Outcomes*, MR-1449-NAVY, Santa Monica, Calif.: RAND Corporation, 2002.

[4] See "Assessment," in Headquarters, Department of the Army, *The Operations Process*, Field Manual Interim (FMI) 5-0.1, Washington, D.C., March 2006a (expires March 2008), pp. 5-1 to 5-6.

[5] Headquarters, Department of the Army, *Operations*, FM 3-0, Washington, D.C., June 2001.

ress. In our usage, evaluation consists of comparing an indicator (whether an MOO, MOE, or MOP) against the criteria of success to judge success or progress.

- *Operational objective:* Desired end state at the operational level; its metric is the MOO.
- *AO objective:* Desired end state at the tactical level for a given AO; its metric is the AO-level MOO.
- *AO effect:* Desired end state or effect for an AO task; its metric is the MOE, which measures whether the desired effect from a task has been achieved.
- *AO task:* Task the force as a whole must perform, or conditions the force must meet, to achieve the end state and AO objective.
- *Measure of outcome (MOO):* Metric used to assess whether the desired operational objective or AO objective has been achieved. To determine whether the MOO has been achieved, one establishes an indicator and criterion of success and measures actual conditions against the desired AO-level end state as operationalized by the indicator and criterion of success. If the indicator, when compared with the criterion, suggests that the desired end state has been achieved, then the campaign objective or AO objective is declared to have been achieved as well.
- *Measure of effectiveness (MOE):* Criterion used to assess changes in system behavior, capability, or operational environment that is tied to measuring the attainment of an end state, achievement of an objective, or creation of an effect (FMI 5-0.1, p. 1-15). MOEs are used to measure results achieved in the overall mission and execution of assigned tasks. Measures of effectiveness are a prerequisite to the performance of combat assessment. (JP 3-60) In terms of IO, MOEs determine whether IO actions being executed are having the desired effect toward mission accomplishment— i.e., the attainment of end states and objectives. MOEs measure battle space results. (JP 3-13) In our usage, MOEs answer the question, Are tasks achieving effects that move us toward our desired end state, or are additional or alternative tasks or actions required? At the tactical level, for each key task, one measures actual effects against the desired criterion of success. If the indi-

cator for assessing the end state for that task meets or exceeds the criterion, then the task is declared to have been effective or successful and the AO effect achieved.[6]

- *Measure of performance (MOP):* Criterion used to assess friendly actions ("inputs"); it is tied to measuring task accomplishment. MOPs answer the question, Was the task or action actually performed as the commander intended? At the unit or system level, for each key task, one measures the performance of friendly efforts against the commander's desired level of task accomplishment. If the desired level of task accomplishment has been achieved, then the performance is declared to have been satisfactory.

An Illustrative Application

The desired AO-level end state captured in the AO objective is the result of the contributions of multiple effects, tasks, and capabilities. MOEs and MOPs, however, with their close focus on tasks, allow IO performance to be tracked more closely. As described above, the MOE focuses on the *effect* produced by one or more related tasks, while the MOP focuses on the *performance* of the capabilities on their assigned tasks.

Consider the example in Figure C.4. In this case, the local AO objective is to persuade 50 percent of enemy forces to capitulate and to defeat 80 percent of the forces that do not capitulate. For purposes of illustration, we have identified two local effects that might be required to support the local AO objective (in practice, there probably would be others as well): (1) move unopposed through Objective Orange, and (2) persuade 50 percent of the enemy forces to capitulate. Note that the first of these requires a combination of IO and other capabilities, whereas the second requires IO capabilities alone.[7]

---

[6]   FMI 5-01 (Headquarters, Department of the Army, 2006a); JP 3-60 (DoD, *Joint Targeting*, JP 3-60, Washington, D.C., April 13, 2007); and JP 3-13 (DoD, 2006a).

[7]   In practice, maneuver or fires probably also would be used in combination with IO or influence operations to compel capitulation.

Figure C.4
Example of IO and Other Tasks

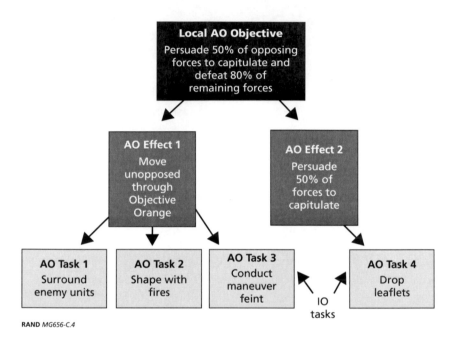

RAND *MG656-C.4*

In this example, we use MOEs to track the degree to which the two AO effects were achieved and MOPs to track the degree to which key tasks were performed to the desired level.

In Figure C.5, the AO objective, now quantified as an MOO, is specified as follows: "Persuade 50 percent of units to capitulate and defeat 80 percent of the remaining units in the AO." Two indicators are used (units capitulating and forces defeated), so two criteria of success (50 percent and 80 percent) are used. Multiple intelligence and other data sources also probably would be required to assess the local AO outcomes, and could include combat assessment data from various intelligence channels, as well as reporting by operators, staff liaisons, and others.

The addition of quantitative indicators and a criterion of success provides benefits throughout the planning and assessment process. Planners now have a clearer idea of the level of resources that will be needed to meet the local AO objectives, and how much effort should

Figure C.5
Example: Measure of Outcome for AO Objective

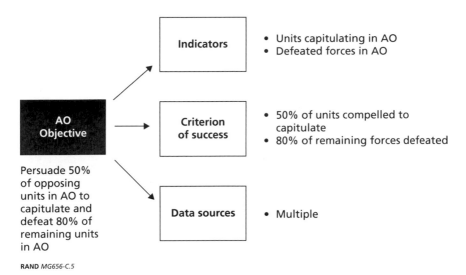

RAND *MG656-C.5*

be directed to persuasion or defeat of opposing units. Operators also know who to target, while assessors have a clearly specified means of measuring "success" in a way that is in line with their commander's objectives.

Drilling down to the next level, Figure C.6 provides an example of an MOE related to the previous example. This MOE measures the AO effect, which in this case is to "persuade 50 percent of opposing units in the AO to capitulate," one component of the overall local AO objective described earlier.

In this example, the desired AO effect is indicated by a single criterion of success for whether the tasks have achieved an AO effect: 50 percent of enemy units capitulating. Usually the evidence for MOEs will be some form of intelligence gathered in support of combat assessment, but it also could be from reporting by operators, liaisons, or others.

Figure C.7 provides an example of an MOP. This MOP focuses on the performance of IO elements and related activities assigned to support accomplishment of a task. Moreover, and subject to the earlier caveat about establishing cause-and-effect relationships, the MOP for

Figure C.6
Example: Measure of Effectiveness

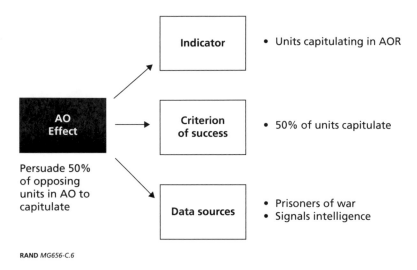

RAND *MG656-C.6*

the AO task that is supposed to produce the AO effect can be used in conjunction with the MOE to assess IO's contribution.

As defined earlier, MOPs traditionally are used to measure inputs—i.e., the task performance of IO elements and related activities in performing their assigned task. MOPs provide the commander and his staff with diagnostic information that illuminates whether these activities were executed or whether they were executed at the desired performance levels.

In this example, the AO task is to "send capitulation message to 75 percent of opposing units," and the criterion is successful delivery of the message to 75 percent of the enemy units.[8] The indicator for the MOP is whether IO elements successfully executed their missions to deliver the message to at least 75 percent of the enemy units—i.e., whether aircraft actually executed their missions and dropped leaflets on enough enemy unit positions to have reasonably reached 75 percent of the units, or whether IO elements actually broadcast the message over known enemy communications frequencies with sufficient power

---

[8]   As a practical matter, it may be necessary to send the message to a larger percentage of units to achieve the 75 percent criterion of success.

Figure C.7
Example: Measure of Performance

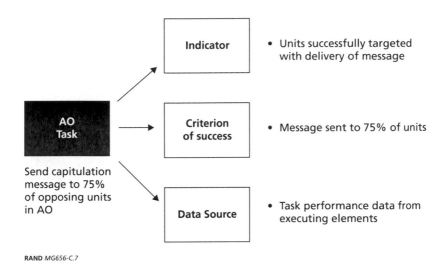

to create a broadcast footprint that covered the presumed position of 75 percent or more of the adversary units.

It is important to note that MOPs can focus either on input measures, such as "units successfully targeted w/delivery of messages," or on output measures, such as "units receiving the message." The difference is that MOPs focused on input measures must be based on performance-related execution data, and MOPS focused on output measures have to be based on some sort of intelligence collection or other combat-assessment data collection activity.

As an example of this difference, consider the following. Leaflets might be successfully delivered on an enemy unit's position but then ignored by members of the unit. To measure the performance of IO elements' activities (or other assets), one would have to know whether there is evidence that the targets actually received (e.g., picked up, listened to, or read) the communication. In this output-focused view, it might be better to judge performance not on the basis of execution data but, rather, on the basis of signal intelligence data, imagery, or prisoner-of-war interrogation reports that would provide anecdotal evidence that the message was received.

However, a focus on output-based MOPs, if it came at the expense of data on input activities, could deprive commanders and their staffs of diagnostics on plan execution that are needed to assess performance. Thus, a combination of input-based MOPs and MOEs might be of higher value than a combination of output-based MOPs and MOEs, if the latter came at the cost of critical execution data. Moreover, requiring collection of output-based MOPs and MOEs could be somewhat redundant in that the MOEs ultimately capture results of greatest interest, and it could create additional collection burdens for intelligence and other sources of combat assessment data.

In the end, the issue seems to reduce to the matter of whether battle staffs can measure inputs, outputs, *and* effects without creating substantial additional data collection and assessment burdens. To the extent that they can have all three, all the better; but if not, hard decisions may be required in choosing input- or output-based MOPs.[9]

Step-by-Step Approach

There are nine steps in the metrics-based planning and assessment approach for developing metrics for IO activities. As will be demonstrated, these nine steps appear to fit reasonably well within existing doctrine for planning, preparation, execution, and assessment of IO, as described in Figure C.8.

Planning Phase
Four of the steps in the metrics-based planning and assessment process are completed during the planning phase, which begins with receipt of a mission and concludes with orders production.[10]

---

[9]   For a good discussion of MOEs and MOPs in assessment, see pp. 5-5 to 5-6 of FMI 5-0.1 (Headquarters, Department of the Army, 2006a).

[10]   A good description of the detailed steps involved in IO planning can be found in Chapter Five of FM 3-13 (Headquarters, Department of the Army, 2003c). The description here highlights only the elements that relate directly to RAND's metrics-based planning and assessment process.

Figure C.8
The Planning, Preparation, Execution, and Assessment Cycle

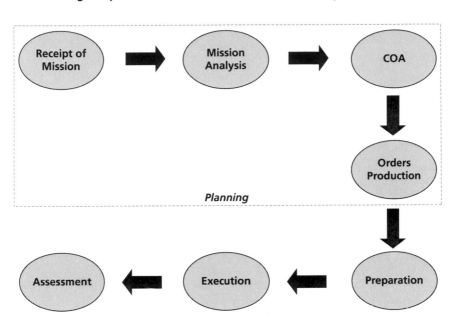

RAND *MG656-C.8*

Step 1. Establish Measures of Outcome. The metrics-based planning and assessment process assumes that campaign- or operational-level objectives are first established by the combined or joint force commander, and that these objectives are translated into MOOs that can be used to judge the campaign's progress. The operational objectives and MOOs are communicated to subordinate units along with the higher headquarters' specified and essential tasks.

Upon receipt, the subordinate unit's G-5 staff establishes MOOs for its AO by translating the guidance from higher echelons and the AO commander into clear, measurable, AO objectives with specific goals appropriate for that echelon in the form of AO MOOs. All subsequent performance and effectiveness measures flow from these objectives.

Step 2. Specification of IO Targets and IO Tasks During Mission Analysis. The entire command staff collaboratively begins the mission analysis process, which includes analysis of the higher headquarters' order and identification of specified, implied, and essential tasks;

a review of available IO and other assets; identification of the initial commander's CCIRs, and other activities.[11]

With inputs from the G-7, the G-2 leads the IPB process, which, among other things, characterizes the information dimension of the operating environment, especially as it relates to enemy forces, sympathizers, neutrals, or friendly parties who might be targeted by IO and influence operations.[12] This process leads to the identification of HVTs, which are subsequently reviewed by the G-7 during the course of action (COA) development process, the purpose of which is to identify IO high-priority targets (HPTs) that should be submitted to the targeting team as part of the HPT list.

Step 3. Develop Measures of Effectiveness for IO Tasks. In current doctrine, the G-7 identifies, concurrent with the IPB process, the AO specified, implied, and essential IO tasks in the higher headquarters' operation plan/operation order that must be performed to support mission accomplishment. As suggested in Figure C.3, however, our metrics-based planning process involves a somewhat different ordering.

Also in current doctrine, specified and essential IO tasks for an AO are identified by higher headquarters and communicated to subordinate echelons. For these specified and essential IO tasks, our methodology envisions that higher headquarters also would communicate to subordinate echelons the desired AO effects in terms of specific MOEs, as described earlier. Moreover, we think that rather than focusing on implied AO tasks from higher headquarters, subordinate AOs should first identify implied AO effects and then identify the AO tasks that can produce these effects. In accomplishing this, the G-7 and G-5 develop the relevant MOEs for IO tasks in the AO that effectively will measure changes in the disposition, condition, or behavior of selected IO targets. The G-5 should approve the MOEs for IO tasks to ensure

---

[11]   FM 3-13 (Headquarters, Department of the Army, 2003c, p. 5-8) identifies a total of 17 tasks conducted during mission analysis.

[12]   FM 3-13 (Headquarters, Department of the Army, 2003c, pp. 5-10 to 5-11) provides examples of the sorts of IPB products that are relevant to the conduct of IO and influence operations.

they nest well with metrics being used by higher headquarters and subordinate echelons.

During mission analysis, the G-7 also coordinates with the G-2 and others on a collection plan to provide the needed data for assessing effectiveness. The G-2's involvement is essential, since the G-2 will provide much of the combat assessment data that will be needed to assess the effectiveness of IO tasks against specific targets. IO or other operators also will need to be tapped for data they might provide on targets and their behavior, but in most cases the data on the effectiveness of IO tasks will be provided by others.

Current doctrine envisions that during COA development, the G-7 develops, based on the initial IO mission statement, a distinct IO concept of support, IO objectives, and IO tasks for each COA. During this phase, our methodology suggests a slight modification in that development of AO effects and MOEs would precede identification of the AO tasks expected to be necessary to create these effects (see Figure C.3).

Step 4. Develop Measures of Performance for IO Tasks. Current doctrine envisions that IO cell members will identify IO elements or related activities that can be employed to accomplish each task and thereby achieve IO objectives for each course of action (in our usage, AO effects). This necessarily involves a review of available IO assets at that echelon, as well as identification of assets that may be available at higher headquarters or lower echelons.

This fourth step in our methodology has the G-7, in coordination with the G-3, developing and approving MOPs that can be used to measure task accomplishment for each element assigned to an AO task; MOPs should also be approved by the G-5. For each AO task, a single executing element and a single MOP are envisioned. The G-3 is essential to this task, because the G-3 will track the overall performance of all Blue activities on the battlefield and will collect data on MOPs for all warfighting functions, including IO; the G-7 will focus on tracking IO activities. Once MOPs are established for all warfighting functions, the G-3 can synchronize the operational plan. The G-7 develops a distinct IO concept of support for the COA, as well as IO objectives, effects, tasks, input work sheets, an IO synchronization matrix that can be integrated with the unit's overall synchronization matrix, and IO-related target nominations.

Step 5. Develop Execution Data Collection Plan. It is at this point in the process that the G-7, in concert with the G-3, develops a collection plan to track Blue execution data needed to assess the IO elements' task performance. Ideally, the G-3 would construct an overall collection plan for execution data for all elements, MOPs, and warfighting functions, since such a plan will facilitate monitoring and assessment, especially the ability to isolate the contributions of different warfighting functions.

Step 6. Finalize Effects Data Collection Plan. The reader will recall that the G-7 develops a collection plan for tracking the effects of operations on IO targets during step 3; in step 6, the data collection plan is finalized for all approved courses of action.

Execution Phase

Step 7. Track IO Measures of Effectiveness and Measures of Performance During Execution. As part of the execution of the IO portion of the commander's plan, the G-7 tracks, as part of its running estimate, the indicators that have been established to underwrite IO MOEs; the G-7 also tracks execution data that were identified as indicators for MOPs.

The key benefit of establishing MOPs and MOEs during the planning stage is realized at this point, as all planning efforts were geared toward a successful and clean execution of the battle plan. By establishing MOPs and MOEs during the planning phase and designing a comprehensive collection plan, the commander's staff gains the ability to monitor execution and effects and to thereby manage the commander's information needs more effectively, as well as to respond to the commander's changing priorities. The use of specific performance indicators lets the commander know sooner rather than later whether the plan is being executed as hoped, and the use of effectiveness indicators illuminates whether desired effects are being achieved. Thus, the measurement approach outlined here facilitates a rapid decisionmaking process.

Assessment Phase

Step 8. Evaluate Measures of Effectiveness and Measures of Performance. In this step, G-7, in conjunction with, respectively, the G-2 and G-3, assesses whether the indicators developed for the MOEs and MOPs have reached or exceeded their criteria of effectiveness. Both assessments are conducted with help from IO operators and other staff elements. The G-7 feeds the results of the IO assessment to the G-5, who maintains overall responsibility for assessing the commander's plan.

Step 9. Evaluate Measures of Outcome. Finally, the G-5 provides the commander with an overall assessment of the battle plan's progress in terms of whether the AO objectives and MOOs have met or exceeded the criteria of success established during planning. If the AO objectives have been achieved, and task-level data on IO and other performance and effects have been collected via the MOP and MOE indicators, the G-5 also may have a basis for hypothesizing about the ways IO contributed to the achievement of the outcome relative to the other warfighting functions or external factors.

Implementation Considerations

The purpose of this section is to demonstrate that with only modest changes, existing doctrine should easily accommodate our metrics-based planning and assessment process.

Doctrinal Considerations

A review of available doctrine on the command process indicates that the functional requirements for planning, executing, and assessing IO and influence operations are essentially the same at every echelon above company level.[13] Moreover, our doctrinal review suggests that the basic building blocks for implementing our metrics-based plan-

---

[13] Of the many documents we reviewed, the following were especially helpful: FMI 5-0.1 and FM 3-0 (Headquarters, Department of the Army, 2006a and 2001, respectively), and FM 6-0 (*Mission Command: Command and Control of Army Forces*, Headquarters, Department of the Army, Washington, D.C., August 2003a).

ning and assessment process using the military decisionmaking process (MDMP)—from receipt of a mission, through mission analysis and COA assessment and approval, to production of a plan or order— already are in place.[14] In particular, recent doctrine's emphasis on the integrated planning and execution of lethal and non-lethal fires (including IO and influence operations) as part of combined arms operations appears to provide a sound basis for the incorporation of metrics into planning and assessment.[15]

We also note that existing doctrine provides for a set of work sheets, lists, matrices, and similar tools that capture much of the information required by our methodology (see Figure C.9), and should be easily adapted to provide the necessary logical data architecture to underwrite our metrics-based planning and assessment process.

As Figure C.9 shows, the IPB process identifies HVTs and provides an initial assessment of their condition that can be updated during execution to support combat assessment. HVTs subsequently are screened in the COA development process to identify HPTs captured in an HPT list or a targeting matrix. Blue IO actions against these targets are synchronized via an IO synchronization matrix, and the actual performance of Blue capabilities against these targets during execution is captured in an execution matrix. Finally, an assessment matrix captures combat assessment data for these targets' post-execution conditions. If a condition meets or exceeds the criteria established in the metrics-based planning and assessment methodology, then the desired effect is declared achieved; if not, retargeting follows.

Existing doctrine details the use of all these tools. For example, current Army IO doctrine discusses the IO synchronization matrix,

---

[14] See, for example, the following: FM 3-13 (Headquarters, Department of the Army, 2003c); FM 5-0 (*Army Planning and Orders Production*, Headquarters, Department of the Army, Washington, D.C., January 2005a), pp. 3-1 to 3-60; FMI 5-0.1 (Headquarters, Department of the Army, 2006a); and FMI 3-90.6 (*Heavy Brigade Combat Team*, Headquarters, Department of the Army, Washington, D.C., March 2005b).

[15] See, for example, Headquarters, Department of the Army, *HBCT Fires and Effects Operations*, FMI 3-09.42, Washington, D.C., April 2005c, especially Chapters 6, 7, and 8.

Figure C.9
Notional Data Flow Through Matrixes

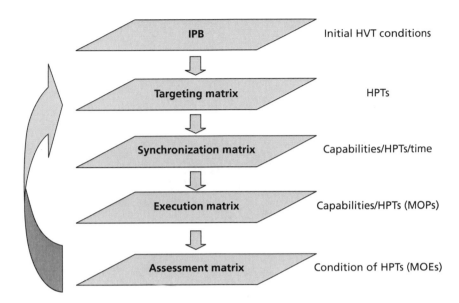

IPB — Initial HVT conditions

Targeting matrix — HPTs

Synchronization matrix — Capabilities/HPTs/time

Execution matrix — Capabilities/HPTs (MOPs)

Assessment matrix — Condition of HPTs (MOEs)

RAND *MG656-C.9*

IO execution matrix, IO assessment matrix, and other tools,[16] and doctrine for heavy brigade combat team (HBCT) fire and effects coordination details the use of the HVT list, HPT list, lethal and non-lethal target/effects synchronization matrixes, and the fire and effects execution matrix. Also, the IO annex of the operation plan or operation order usually includes an IO execution matrix and an IO assessment matrix as appendixes.[17] Thus, there is strong reason to think that these

---

[16]   See, for example, Appendix B of FM 3-13 (Headquarters, Department of the Army, 2003c), which describes the IO input work sheet (p. 5-24) and includes an example of an IO synchronization matrix (p. B-33), IO assessment matrix (pp. 5-25 and B-39 to B-42), and IO execution matrix (p. D-15). The IO input work sheet is a matrix that specifies information on the IO concept of support for a course of action, the IO objective, IO tasks that support the IO objective, criteria of success for each IO task, and information required to assess each IO task.

[17]   FM 3-13 (Headquarters, Department of the Army, 2003c), pp. 5-3 to 5-4 and Appendix D.

existing tools can capture much or all of the data needed to underwrite our metrics-based planning and assessment methodology, and that our methodology, with only modest adjustment, can be implemented within the existing MDMP construct.

Organizational Considerations

We also reviewed documentation on the various organizational elements involved in IO planning, execution, and assessment at the corps, division, and brigade levels. Our review suggests that all organizational elements needed to implement our metrics-based planning and assessment process also are in place, and that implementation of the assessment system described above should therefore be relatively easy for Army organizations from the Army Service Component Command (ASCC) down to the brigade level.

Corps and Division Levels. Our review of corps- and division-relevant doctrinal and organizational information suggests that these echelons have the basic elements needed to implement our metrics-based IO planning and assessment process.

The G-7 is the Assistant Chief of Staff, Information Operations, at both the corps and the division level.[18] As IO Coordinator, the G-7 has the primary responsibility for organizing IO and coordinating IO efforts among the different branches and offices that are also involved in IO. This is primarily with the G-5 (Planning) and through establishment of IO cells. The G-7 will also coordinate with intelligence efforts (G-2) and movement and maneuver efforts (G-3) to ensure integration of the IO plan.[19]

According to the August 2003 release of FM 6-0, the G-7's responsibilities center around organizing current IO operations, coordinating the planning process, targeting, staff planning and supervision, and coordinating staff responsibility. The G-7's IO Division has three branches: a current IO branch, an IO plans branch, and an IO targeting branch. The G-7 also has a special staff of officers dedicated

---

[18] Some brigades have a similar S-7 position.

[19] G-2 (intelligence), G-3 (movement and maneuver, fire support, force protection), G-4 (sustainment), G-5 (plans), G-6 (C2), G-7 (IO), and G-9 (CMO).

to various IO efforts: an electronic warfare officer, a military deception officer, an operations security officer, and a PSYOP officer.

In addition to the G-7, there are other IO-related elements at the division level: a civil affairs battalion, a tactical PSYOP company, a staff judge advocate element, and a fires support element. The civil affairs element sets up the CMO center; it also coordinates with the G-9 (CMO) through a civil affairs planning team. The PSYOP company is in charge of offensive IO; the staff judge advocate element is responsible for providing legal support and advice; and the fires support element controls fires, IO, Army airspace C2, and other elements in the division fire support plan. Fire support, along with movement and maneuvers and force protection, is operated through the G-3.

Brigade Level. Our review of brigade-level doctrinal and organizational information suggests that the brigade also has the capability to successfully employ our metrics-based planning and assessment methodology.

At the brigade level, IO is primarily planned, executed, and assessed in the context of fires and effects operations.[20] To support IO planning, execution, and assessment, the HBCT has an IO officer, a PSYOP staff planner, public affairs and civil affairs sections, reconnaissance elements, tactical HUMINT teams, counterintelligence teams, an electronic warfare specialist, and an organic military intelligence company.[21]

Responsibility for the integration of IO and influence operations into combined arms falls largely to the HBCT effects coordinator:

> The HBCT Effects Coordinator (ECOORD) (formerly the Fire Support Coordinator) is responsible for this staff function via the fires and effects section. He is responsible for all lethal and non-lethal effects planning, coordination, and execution for the HBCT. He advises the commander on the capabilities and

---

[20] See FMI 3-09.42 (Headquarters, Department of the Army, 2005c), especially Chapters 6, 7, and 8.

[21] See FMI 3-90.6 (Headquarters, Department of the Army, 2005b), p. 2-15. For a description of the responsibilities of the S-2, S-3, and others involved in planning, executing, and assessing IO, see pp. 3-13 to 3-14 of this FMI.

employments of fires and effects and is responsible for gaining the commander's guidance for desired effects and their purpose. This section provides support for both current and future operations and is present in both the TAAC and MAIN. The section is composed of a lethal effects cells to include a targeting and counterfire element (fires cell) and a Tactical Air Control party (TACP); and non-lethal cells to include an Information Operations (IO) cell, public affairs cell, civil affairs cell and the HBCT Staff Judge Advocate cell.[22]

As just described, the IO cell of the fire and effects cell, which is headed by an IO coordinator, falls under the non-lethal effects cell along with civil affairs and PSYOP. Insofar as the HBCT includes the necessary specialists and also has mechanisms for coordinating both lethal and non-lethal effects, we judge that the brigade level should have little difficulty adopting our metrics-based planning and assessment methodology.

Figure C.10 shows the original conception of the metrics-based planning and assessment system with the appropriate staff elements at each echelon. It is our recommendation that the Army employ this approach at each echelon with few modifications. When the approach is employed systematically, higher echelons will receive a more precise assessment of battlefield effects and accomplishments and will have a clearer picture of the various contributions of different warfighting functions in executing the battle plan.

Conclusion

Our view is that the new joint IO doctrine pushes IO toward measuring effects and outcomes but does not go quite far enough in outlining a successful approach for building metrics. Army IO doctrine provides a strong doctrinal framework for planning, preparing, executing, and assessing IO, and addresses many of the key conceptual issues related to metrics development, but it does not explicitly provide all links needed

---

[22] FMI 3-90.6 (Headquarters, Department of the Army, 2005b), p. B-8.

Figure C.10
Organizational Backbone to Metrics-Based Planning and Assessment System

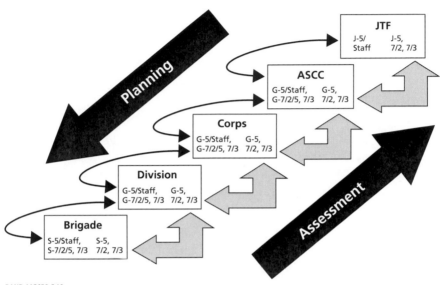

RAND *MG656-C.10*

to connect campaign or operational-level objectives all the way down to executing IO elements and related activities.

A key benefit of establishing MOOs, MOEs, and MOPs during the planning stage is that the approach can help IO operators maintain focus on the "big picture," i.e., the campaign. Clear strategic-, operational-, and tactical-level objectives, along with their associated metrics—MOOs, MOEs, and MOPs—are critical to any metrics-based planning and assessment process. The approach outlined here allows IO planners to link outcomes, effects, tasks, performance, and tasked elements directly to the overall mission objectives, to assess progress on these objectives, and to diagnose reasons for shortfalls.

Thus, our approach for developing metrics for IO and influence operations lays out the critical elements of a sound approach that will assist IO staff planners in identifying additional ways in which IO can support a commander's plan, as well as an effective measurement strategy to assess how IO assets are contributing to the overall mission.

Another benefit of establishing MOOs, MOEs, and MOPs during the planning stage is that it helps manage the commander's information needs and changing priorities and facilitates the rapid decisionmaking process in a proactive, rather than reactive, way. The approach focuses staff attention and data collection efforts on specific indicators and pre-identified data sources.

Lastly, the metrics-based planning and assessment approach can help a commander's staff determine where the plan was deficient in reaching the commander's goals. If MOOs are not being achieved, the G-5 can examine data to diagnose why this might be the case. If the G-5 determines that the MOEs set by the commander are indeed being measured accurately but the intended effects of the task are not being achieved, the staff can identify the impediments to achieving the desired effects by asking some simple but related questions: Which tasks do not appear to be contributing to the effect as intended? Were there shortfalls in the performance of IO and elements and activities on their tasks, as measured by MOPs? Which assets are failing to achieve the desired level of performance?

Moreover, the basic approach outlined in this appendix fits well with current planning, execution, and assessment functions and strengthens existing coordination mechanisms between G-2, G-3, G-5, and G-7. In this approach, MOEs and MOPs are established early and in coordination with all staff elements during the planning stage. This helps the commander to harmonize the efforts of all staff elements at echelons above and below, facilitating the close coordination process essential in modern combat operations. Furthermore, because the approach outlined here incorporates existing staff elements and tasks, it can be implemented fairly easily at every echelon.

Assessment of Expected Utility Modeling for Influence Operations

An earlier study identified agent-based rational choice, or expected utility, modeling as a potentially useful supporting tool for strategy development and planning of influence operations.[1] One of the tasks of the present study was to provide additional information on how these models might be employed. This appendix provides the results of that effort.

Introduction and Overview

Of significant interest to analysts supporting influence operations is how influence operations, either singly or in concert with other policy actions, might influence a target government's—or another stakeholder's—decisionmaking.

The use of expected utility models can help to narrow potential target audiences to stakeholders who have the most influence on the outcome or who will be the easiest to influence. It can illuminate the underlying interest group relationships and dynamics—bargaining, coercion, coalition building, and so on—that influence adversary decisionmaking and interest group politics. Finally, expected utility modeling can bound the range of likely outcomes for a policy dispute

---

[1] See Eric V. Larson, Derek Eaton, Brian Nichiporuk, and Thomas S. Szayna, *Assessing Irregular Warfare: A Framework for Intelligence Analysis*, MG-668-A, Santa Monica, Calif.: RAND Corporation, 2009; and Eric V. Larson et al., forthcoming.

and assist in illuminating strategies that can influence the outcomes of political deliberations in ways that favor U.S. aims.

Of the tools that we identified in our FY 2005 review of methodologies for TRADOC, expected utility modeling is one of the most mature and robust of the approaches we think should be employed in influence strategy development. In our judgment, it is a proven forecasting technique suitable for operational use for influence strategy development.

The expected utility model of principal interest originally was developed by the political scientist Bruce Bueno de Mesquita in the late 1970s and early 1980s.[2] A commercial version of the model, called Policon, was marketed through Data Resources Incorporated and used by the Central Intelligence Agency (CIA) under contract from 1982 to 1986, and an in-house CIA version of the model, called FACTIONS, subsequently was developed by the Directorate of Science and Technology's Office of Research and Development. Some academic versions

---

[2]   The model was adapted to two main applications: international conflict and political forecasting. On applications to interstate conflict, see, for example, Bruce Bueno de Mesquita's "An Expected Utility Theory of International Conflict: An Exploratory Study," *American Political Science Review*, December 1980, pp. 917–931; *The War Trap*, New Haven, Conn.: Yale University Press, 1981; "The War Trap Revisited: A Revised Expected Utility Model," *American Political Science Review*, Vol. 79, No. 1, March 1985; (with David Lalman) *War and Reason*, New Haven, Conn.: Yale University Press, 1992; and *Principles of International Politics: People's Power, Preferences, and Perceptions*, Washington, D.C.: Congressional Quarterly Press, 2000.

On applications to coalition politics, see Bruce Bueno de Mesquita's "Forecasting Policy Decisions: An Expected Utility Approach to Post-Khomeini Iran," *PS: Political Science and Politics*, Vol. 17, No. 2, Spring 1984, pp. 226–236; (with David Newman and Alvin Rabushka), *Forecasting Political Events: The Future of Hong Kong*, New Haven, Conn.: Yale University Press, 1985; (with David Newman and Alvin Rabushka), *Red Flag Over Hong Kong*, Chatham, N.J.: Chatham House, 1996; (with Frans Stokman, eds.), *European Community Decision Making: Models, Applications, and Comparisons*, New Haven, Conn.: Yale University Press, 1994.

of the model are available,[3] and several commercial ventures currently market the tool as part of their consultancies.[4]

The input data required by the model are relatively straightforward:

- identification of the political stakeholder groups within or outside the country that may seek to influence the policy
- a specified range of policy alternatives that encompass all stakeholder groups' preferred outcomes[5]
- the policy preference of each group on the issue
- estimates of the relative political, economic, or military capabilities that each group may employ to influence the policy decision
- estimates of the importance (salience) each group attaches to the issue, signifying the group's willingness to expend political capital to influence policy outcomes.[6]

The model endogenously calculates each stakeholder group's risk orientation, basic orientation toward other actors, estimates of potential

---

[3]   A Web-based version for instructors and students using Bueno de Mesquita's textbook on international politics is available, as of August 31, 2007, at Congressional Quarterly Press's Web site, www.cqpress.com; and a version of the model called EUGene (Expected Utility Generation and Data Management Program), which was developed by D. Scott Bennett and Allan C. Stam III, is available, as of August 31, 2007, at www.eugenesoftware.org for the analysis of datasets dealing with international conflict and cooperation.

[4]   Different commercial versions of the model are in use or marketed by Decision Insights, Inc. (formerly Policon), Policy Futures LLP, and the Sentia Group.

[5]   The model assumes that issues are uni-dimensional, such that preferences can be represented on a line segment, and that preferences are single peaked, so that the associated utilities for potential outcomes diminish steadily the farther in Euclidean distance a possible settlement is from a player's preferred outcome. Each actor's risk orientation is estimated endogenously from his position relative to the predicted outcome, such that actors may perceive the same situation very differently.

[6]   Bueno de Mesquita, 1984. An actor's salience, which can range between 0 and 1, discounts an actor's capabilities. For example, an actor whose capabilities are judged to be 100 and whose salience is 1.0 has an effective capability on the issue of 100, whereas an actor whose capabilities are 100 but whose salience is only 0.5 has an effective capability on the issue of 50.

gains or losses in utility from alternative bargains that might be struck with other actors, and what proposals will be offered and accepted.[7]

The intellectual foundations of expected utility models are found in theoretical work on social choice, spatial voting, and game theory.[8] In their typical contemporary form, expected utility models really consist of two distinct models: first, a weighted spatial voting model forecasts an outcome using whatever voting rule has been specified;[9] second, the model simulates the actual bargaining between actors or groups that are seeking to influence the outcome of the policy issue leading to the predicted outcome.[10] Thus, using a small number of inputs—only three estimates per stakeholder group—the model forecasts an outcome on the policy issue and simulates the interactions between actors—including bargaining and threats—expected to be part of the process leading to that outcome. Given a base forecast, one can then explore policy changes—changes in a U.S. position or salience, for

---

[7]   Actors are assumed to trade off political security and policy gains, with the model inferring each actor's risk orientation—risk acceptance or aversion—from his position relative to the forecast outcome; risk aversion is deemed to increase the closer one is to the forecast outcome, while risk acceptance is deemed to increase the further one is from the forecast outcome.

[8]   Among the foundations are Duncan Black's median voter theorem, William Riker's work on political coalitions, and Jeffrey Banks's theorem about the monotonicity between certain expectations and the escalation of political disputes. See Duncan Black, *The Theory of Committees and Elections*, Cambridge, UK: Cambridge University Press, 1958; William H. Riker, *The Theory of Political Coalitions*, New Haven, Conn.: Yale University Press, 1962; and Jeffrey S. Banks, "Equilibrium Behavior in Crisis Bargaining Games," *American Journal of Political Science*, Vol. 34, 1990, pp. 599–614.

[9]   Most commonly, this is the Condorcet winner, which occupies the median voter position, where actors are weighted by their effective political power. But it also can take the form of qualified majority voting, majority voting with veto, or other voting rules.

[10]   Although the specifics sometimes vary somewhat because of the model's evolution, descriptions of the logical foundations and underlying equations of expected utility models can be found in Bueno de Mesquita, Newman, and Rabushka, 1985, pp. 11–54; Bueno de Mesquita, Newman, and Rabushka, 1996, pp. 165–186; Bueno de Mesquita and Stokman, 1994, pp. 71–104; Bueno de Mesquita, "A Decision Making Model: Its Structure and Form," *International Interactions*, Vol. 23, 1997, pp. 235–266; Bueno de Mesquita, *Predicting Politics*, Columbus, Ohio: Ohio State University Press, 2002, pp. 50–77; and Jacek Kugler, Mark Abdollahian, and Ronald Tammen, "Forecasting Complex Political and Military Events: The Application of Expected Utility to Crisis Situations," Technical Appendix, undated.

example, or in another actor's position or salience—that can shift the political outcome in favorable ways.

Expected utility modeling has been used extensively within the U.S. intelligence community[11] and by academic scholars to forecast political outcomes on various policy issues. It has developed an impressive track record in accurately predicting political outcomes:

- Bruce Bueno de Mesquita (1984) reported that "around 90 percent of the real time forecasts based on this model have proven correct both with respect to the predicted policy decisions and the circumstances surrounding those decisions."[12]
- Stanley A. Feder (2002) reported that during his career at the CIA, he used the voting model on more than 1,200 issues dealing with more than 75 countries.[13] He also reported that in a sample of 80 issues involving more than a score of countries, the voting model alone was accurate almost 90 percent of the time. A 1993 analysis of the likely Italian budget deficit forecast a deficit of 70 trillion lira, within one percent of the deficit that ultimately was approved by the Italian government.[14]
- A. F. K. Organski and S. Eldersveld (1994) evaluated real-time forecasts on 21 policy decisions in the European Community, and concluded that "the probability that the predicted outcome was what indeed occurred was an astounding 97 percent."[15]

---

[11] Stanley Feder reports that the version of the expected utility model in use within the CIA's Directorate of Intelligence had been used in well over 1,000 policy issues of interest to the Directorate. See Stanley A. Feder, "FACTIONS and Policon: New Ways to Analyze Politics," in H. Bradford Westerfield, ed., *Inside CIA's Private World: Declassified Articles from the Agency's Internal Journal, 1955-1992*, New Haven, Conn.: Yale University Press, 1995, pp. 274–292.

[12] Bueno de Mesquita, 1984, p. 233.

[13] Stanley A. Feder, "Forecasting for Policy Making in the Post-Cold War Period," *Annual Review of Political Science*, 2002, pp. 111–125.

[14] Stanley A. Feder, 1995, pp. 274–292.

[15] A. F. K. Organski and S. Eldersveld, "Modeling the EC," in Bruce Bueno de Mesquita and Frans N. Stokman, eds., *European Community Decision Making: Models, Applications, and Comparisons*, New Haven, Conn.: Yale University Press, 1994, pp. 229–242.

- More recently, in late 2002, Jacek Kugler and his associates used expected utility modeling as the basis for an analysis of the imminent war between the United States and Iraq and predicted a U.S. victory and the collapse of Iraqi conventional forces. They also suggested, however, the likely persistence of an insurgency in the aftermath of major combat operations, and an updated analysis immediately after the conclusion of combat operations forecast the break between Ahmed Chalabi, a U.S. protégé, and the U.S. government, among other developments.[16]

Analyses based on expected utility modeling can provide analysts with an understanding of the political dynamics behind likely political outcomes, generate forecasts of the most likely political outcomes under different circumstances and policies, and illuminate the sorts of policy changes that can influence those outcomes in favorable ways. More importantly for present purposes, expected utility modeling is an ideal environment for identifying the most important stakeholder groups, exploring alternative influence strategies for those groups, and estimating the resources that may be required to achieve a desired outcome—all essential to effective influence operations.

Basic Procedure

Four steps are involved in the expected utility approach: data collection, data entry into the model, simulation, and sensitivity and other analyses. The modest data requirements of the model are apparent in

---

[16] See Brian Efird and Jacek Kugler, "Assessing the Stability of Saddam Hussein's Regime," prepared at Claremont Graduate University for the Center for Technology and National Security Policy, National Defense University, 2003; Jacek Kugler and Ronald L. Tammen, "War Initiation and Termination—Exploring the Asian Challenge in the Context of the Iraq War," paper presented to annual meeting of the American Political Science Association, Philadelphia, Pa., August 24–30, 2003; and Michael Baranick, Mark Abdollahian, Brian Efird, and Jacek Kugler, "Stability and Regime Change in Iraq: An Agent Based Modeling Approach," presentation to Military Operations Research Society, Summer 2004. One of the present authors (Larson) provided expert data on Iraqi stakeholder groups that were used in the expected utility modeling simulations and analyses.

Figure D.1, which provides the data used in a Fall 2002 forecast of the outcome of war in Iraq.

Country or regional subject matter experts typically provide the data because they are capable of (a) identifying key stakeholders for an issue, whether groups or individuals; (b) estimating each stakeholder's

Figure D.1
Illustrative Data for Expected Utility Forecast, August 2002

Stakeholder	Resources	Position	Salience	Group	Group Influence
Saddam Hussein	100	100	99	Iraq	100
Uday	10	95	80	Iraq	100
Qusay	15	95	98	Iraq	100
Ali Hasan Al-Majid (cousin)	2	95	95	Iraq	100
Abid Hamid Hamud—Albu Nassar clan	30	95	95	Iraq	100
Tikriti mafia	30	90	90	Iraq	100
Baath party (upper level)	10	75	70	Iraq	100
Baath party (lover level)	5	40	30	Iraq	100
Military—Internal Security Services	15	95	90	Iraq	100
Military—Special Security Organization	10	95	95	Iraq	100
Military—Special Republican Guard	15	90	90	Iraq	100
Military—Republican Guard	25	80	80	Iraq	100
Military—Regular Army Officers	10	60	50	Iraq	100
Military—Regular Army Recruit	5	50	25	Iraq	100
Shiites—Pro-Saddam (secular and urban)	10	80	70	Iraq	100
Shiites—Anti-Saddam (mostly southern)	30	15	80	Iraq	100
Shiites—religious leaders	10	10	40	Iraq	100
Sunnis—middle class (urban)	15	55	40	Iraq	100
Sunnis—masses (urban)	30	50	30	Iraq	100
Sunnis—clan leaders (outer circle)	15	85	80	Iraq	100
Kurds—KDP (Barzani)	15	20	80	Iraq	100
Kurds—PUK (Talabani)	10	15	90	Iraq	100
USA	140	0	90	Foreign	120
United Kingdom	15	10	75	Foreign	120
United Nations (Kofi)	5	60	50	Foriegn	120
Russia	10	60	60	Foreign	120
China	7.5	70	25	Foriegn	120
France	10	70	60	Foreign	120
Saudi Arabia	15	30	90	Foreign	120
Turkey	25	80	65	Foreign	120
Jordan	12.5	60	50	Foreign	120
Syria	10	50	30	Foreign	120
Iran	25	50	70	Foreign	120

SOURCE: Baranick et al., 2004, slide 17.
RAND *MG656-D.1*

relative capabilities; (c) estimating the relative salience of the issue for each stakeholder; and (d) identifying stakeholders' preferred outcomes on an issue continuum. Given that analysts typically generally agree on the input data, data collection can either be accomplished by structured data collection sessions with individual analysts or via group discussion. The result of the data entry phase is a spreadsheet matrix of data providing the position of each stakeholder on a continuum ranging from 0 to 100 (with the most extreme positions connoted by 0 and 100, respectively), as well as estimates of the capabilities and salience of the issue for each stakeholder group (see Figure D.1).

Data are imported into the simulation model, and the analyst runs one or more simulations or sensitivity analyses to understand the most likely outcome under present circumstances and to explore the strategy space and identify influence strategies for promoting different outcomes. For example, the analyst can identify gaps between actors that might be exploited, or can target stakeholder groups for which U.S. leverage is high and only minimal additional U.S. effort—whether in the form of carrots or sticks—might shift the equilibrium outcome in favorable ways.

Implementation Considerations

Over the past several years, Michael Baranick of the National Defense University has been promoting the use within the DoD and intelligence community of an analytic tool called Senturion, which was developed by the Sentia Group.[17] Most recently, in 2008, RAND Arroyo Center personnel began developing an agent-based rational choice, or expected utility, model that offers many of the capabilities of other versions, as well as some additional capabilities, for employment in Army studies. As RAND's model has not been tested or fielded, the following comments relate to Senturion, which appears to be the most popular version of the model in use in the defense community.

---

[17] Senturion is sold by the Sentia Group.

According to our interviews with Sentia Group personnel, Senturion provides all the basic functionality of the earlier generation of the expected utility model that is discussed in the academic literature, as well as a number of new capabilities not available in the earlier model.[18]

Our conversations with Baranick and Sentia Group personnel revealed that the Senturion model already is being employed by a number of regional combatant commands and other DoD organizations.[19] According to the model users with whom we spoke, the model already is being used in a fairly wide range of applications closely related to influence operations, including conflict analysis, analysis of alternative engagement strategies, and identification of stakeholders best situated to carry specific STRATCOMM messages.[20] One user with whom we spoke indicated that the model's most important capability is its ability to illuminate potential strategies and assist in identifying key stakeholders that should be targeted for influence efforts, as well as its ability to support sensitivity analyses.

The favorable comments of the users we interviewed suggest that it should be fairly easy for Army and other defense organizations to incorporate the model as a support tool for developing influence strategies. DoD users view the data requirements for the model as modest and sensible and think that the needed data usually can be provided

---

[18]   Sentia Group personnel stressed that Senturion contains a number of significant intellectual advances over the expected utility model documented in the scholarly literature. As one example, they cited Senturion's ability to simultaneously consider trade-offs across two policy issues.

[19]   For example, Sentia Group literature mentions an analysis of the likely circumstances of the January 2005 Iraqi elections for the DoD, and the leadership transition in Palestine after the death of Arafat for the Defense Intelligence Agency. See Sentia Group, "Senturion Capabilities Overview," undated.

[20]   In one regional combatant command, the model's ability to forecast the behavior of stakeholder groups in response to U.S. actions has led to the model's consideration for another novel application: exercise support. In this application, war gamers choose the COAs they want entered into the model and then respond to the resulting equilibrium situation and the specific moves forecast for other stakeholders.

by country or regional subject matter experts.[21] Experience with the model suggests that most subject matter experts tend to agree on the basic inputs to the model but tend to arrive at quite divergent forecasts without the aid of the model. Moreover, where analysts disagree, one can conduct sensitivity analyses to ascertain whether the differences actually result in different predicted outcomes—in cases where forecasts converge in spite of these differences, the forecast is robust; in cases where forecasts diverge, the analysts can explore in greater detail the reasons for the divergence and, in many cases, following a more detailed analysis, narrow these differences.[22] Finally, the outputs typically were described as intuitive and helpful.

Our sense is that the best place to situate expected utility modeling in field organizations would be in the plans division (J-5/G-5) at the regional combatant command or joint force command level, although we could not rule out that the modeling might be profitably employed within a large corps-level organization such as the Multi-National Corps–Iraq that has the mission of conducting a campaign. Alternatively, expected utility-based assessments for influence operations might be conducted on behalf of Army organizations at the NGIC. A joint force command presumably also could draw on analyses conducted or commissioned by the Defense Intelligence Agency or by the regional desks in the Joint Staff.

Conclusion

We were somewhat surprised not just by the active user community we found within the defense establishment for the modeling approach, but also by the significant resonance for our recommendation that the expected utility modeling approach be employed more widely as a tool in strategy development for influence operations. Moreover, based on

---

[21]   One user with whom we spoke confirmed that in his experience, subject matter experts generally agreed on data inputs.

[22]   Put another way, this is not a Delphi approach that forces convergence to some least common denominator.

the ease with which various military organizations have incorporated the tool, we found few reasons to believe that the Army would have any difficulty in effectively incorporating this model into planning of influence operations.

Assessment of Social Network Analysis for Influence Operations

An earlier study identified SNA tools as potentially useful for support of strategy development and planning of influence operations.[1] This appendix describes the results of one task of the present study, which was to provide additional information on how these tools might be employed.

Introduction and Overview

Also of interest in influence operations is understanding the political, military, tribal, religious, patronage, and other networks that are the backbone of decisionmaking and control in a society, and the degree of influence that leaders/nodes in these networks exercise over their followers or subordinates.

As was discussed earlier, understanding these networks is important, because the extent to which planners understand them determines how well influence operations can be targeted at selected, high-payoff nodes that will accomplish objectives much more efficiently than would operations targeted diffusely, across a much larger, more heterogeneous, and less specific audience:

---

[1]   For thoughts on the application of SNA to irregular warfare and influence operations, see Larson et al., 2009; and Larson et al., forthcoming.

- In highly *centralized* political and military systems, the persuasion of individuals at the central node often may be sufficient to compel compliance of the entire network, and their elimination may lead to destabilization or collapse of the system.[2]
- In highly *hierarchical* systems, the successful persuasion of individuals higher in the hierarchy often may be sufficient to compel compliance of those beneath them, either through direct efforts to influence followers or through broader social influence efforts.[3]

Leaving aside the somewhat hyped concept of "netwar" that is in vogue in some quarters, there appears to be growing recognition within the Army and DoD that many phenomena of interest—terrorist groups, the support systems for HVTs such as Saddam Hussein and Osama bin Laden, and even insurgencies—can be characterized and visualized as networks, and that viewing these phenomena through the lens of SNA tools can lead to insights that might otherwise remain elusive.[4]

Our judgment is that there appear to be many SNA tools that will provide a graphical portrayal of networks that may facilitate the analysis of leadership, terrorist, tribal, or other networks, the understanding of which is essential in developing influence strategies.[5]

---

[2]   L. Freeman, "Centrality in Social Networks: Conceptual Clarification," *Social Networks*, Vol. 1, 1979, pp. 215–239.

[3]   For discussions of hierarchy in a network, see D. Krackhardt, J. Blythe, and C. McGrath, "KrackPlot 3.0: An Improved Network Drawing Program," *Connections*, Vol. 17, No. 2, 1994, pp. 53–55; and N. Hummon and T. Fararo, "Actors and Networks as Objects," *Social Networks*, Vol. 17, 1995, pp. 1–26. For a relatively recent review of the social influence literature, see Robert B. Cialdini and Noah J. Goldstein, "Social Influence: Compliance and Conformity," in *Annual Review of Psychology* (Palo Alto, Calif.: Annual Reviews), Vol. 55, 2004, pp. 591–621.

[4]   For example, the final draft of Army FM 3-24, *Counterinsurgency*, includes an appendix that introduces readers to SNA. See Headquarters, Department of the Army, *Counterinsurgency*, Washington, D.C., final draft, June 2006b, pp. E-1 to E-10.

[5]   For example, it has been reported that Saddam's capture was the result of the efforts of an intelligence cell to provide a network mapping of the "pack of cards" of 55 top Iraqi leaders.

There also are a great many SNA tools for analyzing network data and identifying key players by virtue of their degree, closeness, betweenness, information, or eigenvector centrality.[6] But while they may be useful for identifying prominent members of networks using standard measures of centrality, most have very little to say about the influence these members may exercise over others in the network.[7]

Various interesting recent work offers some hope that network analysts may develop additional measures of diffusion, contagion, and influence whose predictive power will be confirmed by empirical analysis.[8] But until then, SNA seems likely to make only a limited con-

---

[6] For a comprehensive list of available SNA computer programs, see International Network for Social Network Analysis, "Computer Programs for Social Network Analysis, last updated December 2005. Other programs in use in the national security, intelligence, and law enforcement communities, such as Analyst's Notebook and the Situational Influence Assessment Model (SIAM), also provide network display and analysis capabilities. SIAM enables the creation of user-specified "influence net models" based on expert judgment about factors that will influence decisions and a forward-propagation algorithm for beliefs about the likelihood of specific factors being true. See, for example, Julie A. Rosen and Wayne L. Smith, "Influence Net Modeling for Strategic Planning: A Structured Approach to Information Operations," *Phalanx*, Vol. 33, No. 4, December 2000.

[7] Analysts have tended to focus on the problem of disrupting terrorist networks, but no conceptual reason precludes the use of some of these approaches in designing influence operations. On the issue of key players, see Stephen P. Borgatti, "The Key Player Problem," in R. Breiger, K. Carley, and P. Pattison, eds., *Dynamic Social Network Modeling and Analysis: Workshop Summary and Papers*, Washington, D.C.: National Academies Press, 2003.

[8] Diffusion within networks is discussed in H. Peyton Young, "Diffusion in Social Networks," Working Paper No. 2, Brookings Institution Center on Social and Economic Dynamics, May 1999; and Dunia Lopez-Pintado, "Diffusion in Complex Social Networks," WP-AD 2004-33, Instituto Valenciano de Investigaciones Economicas, S.A., October 12, 2004. See, for example, the treatment of influence in three presentations given at the 73rd Military Operations Research Society Symposium, U.S. Military Academy, West Point, N.Y., June 21–23, 2005: Richard Avila and Jacob Shapiro, "Social Network Analysis Using Fuzzy Sets"; Clinton R. Clark, Richard F. Deckro, Jeffery D. Weir, and Marcus B. Perry, "Modeling and Analysis of Clandestine Networks"; and J. Todd Hamill, Richard F. Deckro, Victor D. Wiley, and Robert S. Renfro II, *Gains, Losses, and Thresholds of Influence Within a Social Network: A Modeling Approach*. See also David Kempe, Jon Kleinberg, and Eva Tardos, "Maximizing the Spread of Influence Through a Social Network," presentation, KDD-2003: The Ninth Association of Computing Machinery SIGKDD International Conference on Knowledge Discovery and Data Mining, Washington, D.C., August 24–27, 2003; J. Richard Harrison and Glenn R. Carroll, "The Dynamics of Cultural Influence Networks," draft, November 27, 2001.

tribution to influence operations, largely restricted to visualization of network data.

Moreover, most network analyses focus on relatively small and tractable networks, whereas influence operations require an understanding of national-level networks comprising the most influential members of a society and the sway they may hold over their followers, which is a somewhat different problem. The data requirements for this problem are uncertain but will need to be sufficiently simple to support operational use by military operators.

Our discussions with practitioners using SNA tools suggest that these tools typically are used to analyze terrorist groups and the networks surrounding HVTs (e.g., Saddam, bin Laden, Zarqawi).[9] Our conversations revealed that SNA tools are not typically thought of as a way to characterize the larger social structures underpinning authority and influence that must be understood if persuasion in IO and influence operations is to be successful. Thus, to us, the application of SNA tools to influence operations appears to present a largely unrealized opportunity for SNA to assist intelligence analysts and planners in supporting influence operations.

Basic Procedure

In its most basic form, the data used by SNA tools are lists of nodes and the links between nodes in the form of a head, a link, and a tail:

HEAD1	TAIL1	LINK1
HEAD2	TAIL2	LINK2
HEAD3	TAIL3	LINK3

---

[9]   See, for example, Marc Sageman, *Understanding Terror Networks*, Philadelphia, Pa.: University of Pennsylvania Press, 2004; and Valdis Krebs, "Connecting the Dots: Tracking Two Identified Terrorists," 2002 (with updates through 2007), and "Social Network Analysis of the 9-11 Terrorist Network," 2006.

Most SNA tools can import these data from commercial spreadsheet or tab-delimited text files.[10]

Many packages also provide the user with a capability for customizing icons, colors, line types, and other characteristics of nodes and arcs. Figure E.1 is an example of input data from the Pajek program's tutorial page; it illustrates that package's capabilities for customization of nodes ("vertices," in the example), and links ("edges," which are undirected, and "arcs," which are directed edges). As should be clear from this example, readily available software tools provide significant capabilities for visualization of networks and for tailoring those visualizations to highlight the most salient relationships.

Figure E.1
Illustration of Pajek Network Customization

```
*Vertices 3
1 "Doc1" 0.0 0.0 0.0 ic Green bc Brown
2 "Doc2" 0.0 0.0 0.0 ic Green bc Brown
3 "Doc3" 0.0 0.0 0.0 ic Green bc Brown
*Arcs
1 2 3 c Green
2 3 5 c Black
*Edges
1 3 4 c Green
```

Herein there are 3 vertices—Doc1, Doc2 and Doc3—denoted by numbers 1, 2 and 3. The (fill) color of these nodes is Green and the border color is Brown. The initial layout location of the nodes is (0,0,0). Note that the (x,y,z) values can be changed interactively after drawing.

There are two arcs (directed edges). The first goes from node 1 (Doc1) to node 2 (Doc2) with a weight of 3 and in color Green.

For edges, there is one from node 1 (Doc1) to node 3 (Doc3) of weight of 4, and in Green color.

SOURCE: "Pajek Tutorial," undated.
RAND *MG656-E.1*

---

[10] For example, for another study we used an SNA software tool called Pajek ("Networks / Pajek: Program for Large Network Analysis," updated January 1, 2009).

Implementation Considerations

Our research suggests that SNA tools have become ubiquitous within the Army and other military organizations. This, along with the ease of use and ready availability of various software packages appropriate for conducting link and node analysis, leads us to think that Army organizations should have little difficulty incorporating SNA approaches into IPB and planning for influence operations.

Two SNA tools already in use within the Army are Analyst's Notebook and the Counterintelligence Human Intelligence Management System (CHMS). The Analyst's Notebook, developed by I2 systems, enables the Army Counterintelligence Center (ACIC) analysts to prepare and share link-analysis charts;[11] Analyst's Notebook also is being widely used by Army military intelligence personnel in Iraq.[12]

The other SNA tool being used, CHMS, provides counterintelligence personnel at the battalion level and up with link and node analysis capabilities.[13] However, the principal application of this software appears to be for mapping connections within terrorist and insurgent groups, the networks of HVTs, and similar problems. It does not appear to be a tool used in influence operations.

There are, no doubt, other tools in use within the Army and military organizations that provide SNA capabilities, and we would expect that most of these could easily be adapted to representing government, tribal, religious, or other leadership networks in support of influence strategy development.

---

[11] According to Charles E. Harlan in "Developing a Predictive Capability in the Counterintelligence Integrated Analysis Center," *Military Intelligence Professional Bulletin*, January–March 2005: "The [Counterintelligence Integrated Analysis Center] uses Analyst's Notebook to identify links between known or suspected terrorists, their activities, phone numbers, locations, and their associations with other persons, events, or groups. Analyst's Notebook charts are increasingly being used in the ACIC Terrorism Summary (ATS) to help readers understand linkages in the information provided."

[12] See "Computer-Sleuthing Aids Troops in Iraq," *CNN.com*, December 23, 2003.

[13] Author conversation with Larry Schneider, of the Army Science Board.

Conclusion

The research that we conducted, described in this appendix, supports the conclusion that SNA tools may be useful in the visualization of authority and influence structures and other influence operations-related networks. Nevertheless, we think that additional intellectual effort will be needed to extend the SNA toolkit to represent influence within a network and other phenomena of central interest in influence operations.

References

Avila, Richard, and Jacob Shapiro, "Social Network Analysis Using Fuzzy Sets," presentation, 73rd Military Operations Research Society Symposium, U.S. Military Academy, West Point, N.Y., June 21–23, 2005.

Baker, Ralph O., "The Decisive Weapon: A Brigade Combat Team Commander's Perspective on Information Operations," *Military Review*, May–June 2006, pp. 13–32.

Banks, Jeffrey S., "Equilibrium Behavior in Crisis Bargaining Games," *American Journal of Political Science*, Vol. 34, 1990, pp. 599–614.

Baranick, Michael, Mark Abdollahian, Brian Efird, and Jacek Kugler, "Stability and Regime Change in Iraq: An Agent Based Modeling Approach," presentation to Military Operations Research Society, Summer 2004.

Black, Duncan, *The Theory of Committees and Elections*, Cambridge, UK: Cambridge University Press, 1958.

Borgatti, Stephen P., "The Key Player Problem," in R. Breiger, K. Carley, and P. Pattison, eds., *Dynamic Social Network Modeling and Analysis: Workshop Summary and Papers*, Washington, D.C.: National Academies Press, 2003. As of September 2, 2007:
www.analytictech.com/borgatti/publications.htm

"British Memo Says Heavy-Handed U.S. Tactics Have Fuelled Opposition in Fallujah, Najaf," *Associated Press*, May 25, 2004.

Bueno de Mesquita, Bruce, "An Expected Utility Theory of International Conflict: An Exploratory Study," *American Political Science Review*, December 1980, pp. 917–931.

———, *The War Trap*, New Haven, Conn.: Yale University Press, 1981.

———, "Forecasting Policy Decisions: An Expected Utility Approach to Post-Khomeini Iran," *PS: Political Science and Politics*, Vol. 17, No. 2, Spring 1984, pp. 226–236.

———, "The War Trap Revisited: A Revised Expected Utility Model," *American Political Science Review*, Vol. 79, No. 1, March 1985.

————, "A Decision Making Model: Its Structure and Form," *International Interactions*, Vol. 23, 1997, pp. 235–266.

————, *Principles of International Politics: People's Power, Preferences, and Perceptions*, Washington, D.C.: Congressional Quarterly Press, 2000.

————, *Predicting Politics*, Columbus, Ohio: Ohio State University Press, 2002.

Bueno de Mesquita, Bruce, and David Lalman, *War and Reason*, New Haven, Conn.: Yale University Press, 1992.

Bueno de Mesquita, Bruce, David Newman, and Alvin Rabushka, *Forecasting Political Events: The Future of Hong Kong*, New Haven, Conn.: Yale University Press, 1985.

————, *Red Flag Over Hong Kong*, Chatham, N.J.: Chatham House, 1996.

Bueno de Mesquita, Bruce, and Frans Stokman, eds., *European Community Decision Making: Models, Applications, and Comparisons*, New Haven, Conn.: Yale University Press, 1994.

CALL—*See* Center for Army Lessons Learned

Center for Army Lessons Learned, *Initial Impressions Report, Operation Iraqi Freedom: Information Operations, Civil Military Operations, Engineer, Combat Service Support*, Report No. 04-13, Center for Army Lessons Learned, Fort Leavenworth, Kan., May 2004 (not available to the public).

————, *Initial Impressions Report, Information Operations: Information Operations, Organization and Pre-Employment Preparations for Information Operations, Integration of Information Operations Into Planning and Operating*, Center for Army Lessons Learned, Fort Leavenworth, Kan., May 2005a (not available to the public).

————, *Tactical Commander's Handbook, Information Operations: Operation Iraqi Freedom (OIF)*, Combined Arms Center, Fort Leavenworth, Kan., May 2005b (not available to the public).

————, "Integration of Information Operations into Planning and Operations, Public Affairs, and the Media; Extract from Center for Army Lessons Learned Initial Impressions Report 05-3, Information Operations," Chapter Seven in *Media Is the Battlefield*, CALL Newsletter No. 07-04, October 2006, p. 51. As of February 1, 2009:
http://usacac.army.mil/cac2/call/docs/09-11/toc.asp

Chiarelli, Peter W., and Patrick R. Michaelis, "Winning the Peace: The Requirement for Full-Spectrum Operations," *Military Review*, July–August 2005, pp. 4–17.

Cialdini, Robert B., and Noah J. Goldstein, "Social Influence: Compliance and Conformity," in *Annual Review of Psychology* (Palo Alto, Calif.: Annual Reviews), Vol. 55, 2004, pp. 591–621.

Clark, Clinton R., Richard F. Deckro, Jeffery D. Weir, and Marcus B. Perry, "Modeling and Analysis of Clandestine Networks" presentation, 73rd Military Operations Research Society Symposium, U.S. Military Academy, West Point, N.Y., June 21–23, 2005.

"Computer-Sleuthing Aids Troops in Iraq," *CNN.com*, December 23, 2003. As of August 29, 2007:
http://www.cnn.com/2003/TECH/ptech/12/23/tracking.rebels.ap/index.html

Cracknall, David, "British Fears on U.S. Tactics Are Leaked," *Sunday Times of London*, May 23, 2004.

DoD—*See* Department of Defense

Department of Defense, *Department of Defense Dictionary of Military and Associated Terms*, JP 1-02, Washington, D.C., April 12, 2001a (as amended through October 17, 2008).

———, *2001 Quadrennial Defense Review Report*, September 30, 2001b. As of February 5, 2009:
http://www.defenselink.mil/pubs/pdfs/qdr2001.pdf

———, *Joint Doctrine for Urban Operations*, JP 3-06, Washington, D.C., September 16, 2002.

———, *Information Operations Roadmap*, October 30, 2003. As of February 5, 2009:
http://www.gwu.edu/~nsarchiv/NSAEBB/NSAEBB177/info_ops_roadmap.pdf

———, *Universal Joint Task List (UJTL)*, CJCSM 3500.04D, Washington D.C., August 1, 2005.

———, *Information Operations*, JP 3-13, Washington, D.C., February 13, 2006a.

———, *Joint Operations Planning*, JP 5-0, Washington, D.C., December 26, 2006b.

———, *Joint Targeting*, JP 3-60, Washington, D.C., April 13, 2007.

Efird, Brian, and Jacek Kugler, "Assessing the Stability of Saddam Hussein's Regime," prepared at Claremont Graduate University for the Center for Technology and National Security Policy, National Defense University, 2003.

Feder, Stanley A., "FACTIONS and Policon: New Ways to Analyze Politics," in H. Bradford Westerfield, ed., *Inside CIA's Private World: Declassified Articles from the Agency's Internal Journal, 1955-1992*, New Haven, Conn.: Yale University Press, 1995, pp. 274–292.

———, "Forecasting for Policy Making in the Post-Cold War Period," *Annual Review of Political Science*, 2002, pp. 111–125.

1st IOC [Information Operations Command] (Land), Field Support Division, "Terminology for IO Effects," in *Tactics, Techniques and Procedures for Operational & Tactical Information Operations Planning*, March 2004.

Freeman, L., "Centrality in Social Networks: Conceptual Clarification," *Social Networks*, Vol. 1, 1979, pp. 215–239.

George, Alexander L., "The Need for Influence Theory and Actor-Specific Behavioral Models of Adversaries," in Barry R. Schneider and Jerrold M. Post, eds., *Know Thy Enemy: Profiles of Adversary Leaders and Their Strategic Cultures*, Maxwell Air Force Base, Ala.: United States Air Force Counterproliferation Center, November 2002, pp. 271–310.

Gilmore, Gary J., "Afghanistan, Iraq Lessons Learned Part of Joint War Game," *Armed Forces Press Service*, March 28, 2006. As of February 5, 2009: http://www.tradoc.army.mil/pao/TNSarchives/March06/032906-1.html

Hamill, J. Todd, Richard F. Deckro, Victor D. Wiley, and Robert S. Renfro II, *Gains, Losses, and Thresholds of Influence Within a Social Network: A Modeling Approach*, presentation, 73rd Military Operations Research Society Symposium, U.S. Military Academy, West Point, N.Y., June 21–23, 2005; also, briefing, U.S. Air Force Institute of Technology, June 14, 2005.

Harlan, Charles E., "Developing a Predictive Capability in the Counterintelligence Integrated Analysis Center," *Military Intelligence Professional Bulletin*, January–March 2005.

Harrison, J. Richard, and Glenn R. Carroll, "The Dynamics of Cultural Influence Networks," draft, November 27, 2001.

Headquarters, Department of the Army, *Operations*, FM 3-0, Washington, D.C., June 2001.

———, *Mission Command: Command and Control of Army Forces*, FM 6-0, Washington, D.C., August 2003a.

———, *The Army Universal Task List*, FM 7-15, Washington D.C., August 2003b.

———, *Information Operations: Doctrine, Tactics, Techniques, and Procedures*, FM 3-13, Washington, D.C., November 2003c.

———, *Army Planning and Orders Production*, FM 5-0, Washington, D.C., January 2005a.

———, *Heavy Brigade Combat Team*, FMI 3-90.6, Washington, D.C., March 2005b.

———, *HBCT Fires and Effects Operations*, FMI 3-09.42, Washington, D.C., April 2005c.

———, *Psychological Operations,* FM 3-05.30 (MCRP 3-40.6), Washington, D.C., April 2005d.

————, *Psychological Operations Leaders Planning Guide*, GTA-33-01-001, Washington, D.C., November 2005e. As of February 5, 2009:
http://www.fas.org/irp/doddir/army/psyopplan.pdf

————, *The Operations Process*, FMI 5-0.1, Washington, D.C., March 2006a (expires March 2008).

————, *Counterinsurgency*, final draft, Washington, D.C., June 2006b.

————, *Counterinsurgency*, FM 3-24, Washington, D.C., December 2006c.

————, *Stability Operations*, FM 3-07, Washington, D.C., October 2008.

Hummon, N., and T. Fararo, "Actors and Networks as Objects," *Social Networks*, Vol. 17, 1995, pp. 1–26.

International Network for Social Network Analysis, "Computer Programs for Social Network Analysis" (last updated December 2005). As of January 2006:
http://www.insna.org/INSNA/soft_inf.html

Kahan, James P., D. Robert Worley, and Cathleen Stasz, *Understanding Commanders' Information Needs*, Santa Monica, Calif.: RAND Corporation, R-3761-1, 2000. As of February 5, 2009:
http://www.rand.org/pubs/reports/R3761-1/

Kahn, Herman, and Izi Man, "War Gaming," P-1167, Santa Monica, Calif.: RAND Corporation, 1957. As of February 5, 2009:
http://www.rand.org/pubs/papers/P1167/

Kempe, David, Jon Kleinberg, and Eva Tardos, "Maximizing the Spread of Influence Through a Social Network," presentation, KDD-2003: The Ninth Association of Computing Machinery SIGKDD International Conference on Knowledge Discovery and Data Mining, Washington, D.C., August 24–27, 2003.

Krackhardt, D., J. Blythe, and C. McGrath, "KrackPlot 3.0: An Improved Network Drawing Program," *Connections*, Vol. 17, No. 2, 1994, pp. 53–55.

Krane, Jim, "U.S. Steps Back from the Brink in Fallujah," Associated Press, May 1, 2004.

Krebs, Valdis, "Connecting the Dots: Tracking Two Identified Terrorists," 2002 (with updates through 2007). As of February 5, 2009:
http://www.orgnet.com/tnet.html

————, "Social Network Analysis of the 9-11 Terrorist Network," 2006. As of February 5, 2009:
http://www.orgnet.com/hijackers.html

Kugler, Jacek, Mark Abdollahian, and Ronald Tammen, "Forecasting Complex Political and Military Events: The Application of Expected Utility to Crisis Situations," Technical Appendix, undated.

Kugler, Jacek, and Ronald L. Tammen, "War Initiation and Termination— Exploring the Asian Challenge in the Context of the Iraq War," paper presented to annual meeting of the American Political Science Association, Philadelphia, Pa., August 24–30, 2003.

Lamb, Christopher J., "Information Operations as a Core Competency," *Joint Forces Quarterly*, Issue 36, December 2004. As of February 5, 2009: http://www.dtic.mil/doctrine/jel/jfq_pubs/1536.pdf

———, *Review of Psychological Operations: Lessons Learned from Recent Operational Experience*, Washington, D.C.: National Defense University Press, September 2005.

Larsen, Stephen C., "Conducting Psychological Operations in Sophisticated Media Environments," master's thesis, U.S. Army Command and General Staff College, Fort Leavenworth, Kan., 1999.

Larson, Eric V., Richard E. Darilek, Daniel Gibran, Brian Nichiporuk, Amy Richardson, Lowell Schwartz, and Cathryn Thurston, MG-654-A, *Foundations of Effective Influence Operations: A Framework for Enhancing Army Capabilities*, Santa Monica, Calif.: RAND Corporation, forthcoming.

Larson, Eric V., Derek Eaton, Brian Nichiporuk, and Thomas S. Szayna, *Assessing Irregular Warfare: A Framework for Intelligence Analysis*, MG-668-A, Santa Monica, Calif.: RAND Corporation, 2009. As of February 5, 2009: http://www.rand.org/pubs/monographs/MG668/

Lawrence, T. E., "The 27 Articles of T. E. Lawrence," *The Arab Bulletin*, August 20, 1917.

Lopez-Pintado, Dunia, "Diffusion in Complex Social Networks," Working Paper WP-AD 2004-33, Instituto Valenciano de Investigaciones Economicas, S.A., October 12, 2004. As of February 5, 2009: http://www.ivie.es/downloads/docs/wpasad/wpasad-2004-33.pdf

Marquis, Jefferson P., Richard E. Darilek, Jasen J. Castillo, Cathryn Quantic Thurston, Anny Wong, Cynthia Huger, Andrea Mejia, Jennifer D. P. Moroney, Brian Nichiporuk, and Brett Steele, *Assessing the Value of Army International Activities*, MG-329-A, Santa Monica, Calif.: RAND Corporation, 2006. As of February 5, 2009: http://www.rand.org/pubs/monographs/MG329/

Metz, Thomas F., Mark W. Garrett, James E. Hutton, and Timothy W. Bush, "Massing Effects in the Information Domain: A Case Study in Aggressive Information Operations," *Military Review*, May–June 2006, pp. 2–12.

"Networks / Pajek: Program for Large Network Analysis," updated January 1, 2009. Available for download as of February 8, 2009: http://vlado.fmf.uni-lj.si/pub/networks/pajek/

Organski, A. F. K., and S. Eldersveld, "Modeling the EC," in Bruce Bueno de Mesquita and Frans N. Stokman, eds., *European Community Decision Making: Models, Applications, and Comparisons*, New Haven, Conn.: Yale University Press, 1994, pp. 229–242.

"Pajek Tutorial," undated. As of February 5, 2009:
http://vw.indiana.edu/tutorials/pajek/

Perry, Walter L. Robert W. Button, Jerome Bracken, Thomas Sullivan, and Jonathan Mitchell, *Measures of Effectiveness for the Information-Age Navy: The Effects of Network-Centric Operations on Combat Outcomes*, MR-1449-NAVY, Santa Monica, Calif.: RAND Corporation, 2002. As of February 5, 2009:
http://www.rand.org/pubs/monograph_reports/MR1449/

Petraeus, David H., "Learning Counterinsurgency: Observations from Soldiering in Iraq," *Military Review*, January–February 2006, pp. 2–12.

Post, Jerrold M., resume, undated. As of February 5, 2009:
http://www.gwu.edu/~icdrm/programs/facultybios/post.pdf

Riker, William H., *The Theory of Political Coalitions*, New Haven, Conn.: Yale University Press, 1962.

Riva, Julia, review of P. Christopher Earley and Soong Ang, *Cultural Intelligence: Individual Interactions Across Cultures* (Palo Alto, Calif.: Stanford University Press, 2003), in *Studies in Intelligence*, Vol. 49, No. 2, 2005.

Rosen, Julie A., and Wayne L. Smith, "Influence Net Modeling for Strategic Planning: A Structured Approach to Information Operations," *Phalanx*, Vol. 33, No. 4, December 2000.

Sageman, Marc, *Understanding Terror Networks*, Philadelphia, Pa.: University of Pennsylvania Press, 2004.

Schreckengost, Gary J., and Gary A. Smith, "IO in SOSO at the Tactical Level; Converting Brigade IO Objectives into Battalion IO Tasks," *Field Artillery*, July–August 2004, pp. 11–15.

Sentia Group, "Senturion Capabilities Overview," undated.

Sicoli, Peter A., *Filling the Information Void: Adapting the Information Operation (IO) Message in Post-Hostility Iraq*, School of Advanced Military Studies, U.S. Army Command and General Staff College, Fort Leavenworth, Kan., May 2005.

SYTEX, Inc., *Introduction to Information Campaign Planning and Execution*, student materials handbook produced for U.S. Army Land Information Warfare Activity, Vienna, Va., 1997.

Teamey, Kyle, and Jonathan Sweet, "Organizing Intelligence for Counterinsurgency," *Military Review*, September–October 2006, pp. 24–29.

Young, H. Peyton, "Diffusion in Social Networks," Working Paper No. 2, Brookings Institution Center on Social and Economic Dynamics, May 1999.